Volume 2 - South Asia & Indo-China

Author: Abu Hassan Jalil

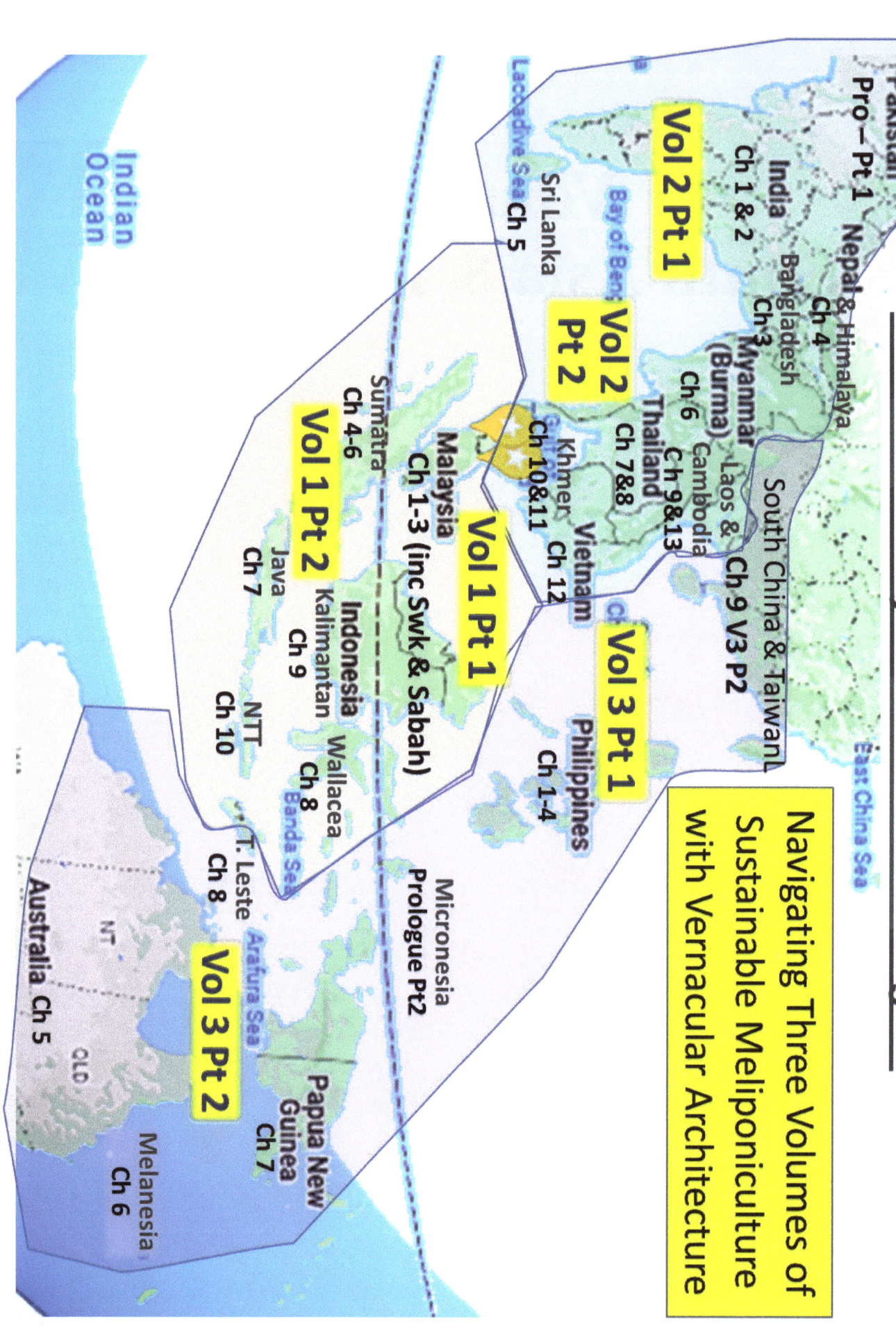

Sustainable Meliponiculture with Vernacular Architecture

Volume 2 - South Asia & Indo-China

Synopsis: Thermal comfort is a grave concern regarding conservation in Meliponiculture. Often enough, enthusiastic new beekeepers are not overly worried about heat dissipation, especially during the summers on the fringes of the tropics.

Heat waves in the tropics may leave beekeepers anxious about the stability of the colony's health and population growth. Chances are the existing colonies may see the population dwindling rather than growing at any rate. We also look at earth tremors, if not earthquakes, in Pakistan Region. Flooding threats in South Asia are very troubling. Recent reports on landslides leave affected beekeepers very distressed.

Addressing these disastrous events and the potential threats is one of the aims of this book. We have scoured the region and examined how different cultures attempt to mitigate their regions' drastic and extreme weather. Out of these examples and data collection, we provide a gallery of designs and relevant bee illustrations whose type locality is mainly in South Asia and Indo-China.

The reader is afforded a choice of constructive possibilities depending on the availability of materials, site topography, and geography of their Meliponary location from the Indian Subcontinent and the Himalayan States, Tenasserim Range and mainland Southeast Asia.

Acknowledgements

The author is very grateful to Dr Shashidhar Viraktamath, Emeritus Scientist, Department of Entomology, University of Agricultural Sciences, GKVK, Bengaluru, India, for his advice and edits in the portion Part 1 – Indian Subcontinent.

We are much obliged to our mentor, Dr. David Ward Roubik of the Smithsonian Tropical Research Institute, Panama, for all their extremely useful and indispensable advice and guidance towards preparing major portions of this book.

The author is very grateful to Dr *Claus Rasmussen*, the Danish entomologist. Department of Agroecology, Aarhus University, Blichers Allé 20, 8830 Tjele, Denmark, for his advice and for sharing the Type Locality of the many Meliponine species in this Indo-Malayan/Australasian region.

We also express gratitude to Dr. Pham Hong Thai, Director of the Center for Bee Research and Tropical Beekeeping at QT MEDICINE from Hanoi, Vietnam, for suggestions in Beekeeping among ethnic cultures in Vietnam.

Appreciation goes to Dr. Myint Thuzar, 3rd Professor at Yezin Agricultural University, Zeyarthiri, Nay Pyi Taw Union Territory, Myanmar, for suggestions on the division of ethnic peoples in Myanmar.

Many thanks to Dr. Anchalee Sawathum, from the Rajamangala University of Technology Thanyaburi, Thailand, for literary contributions on Stingless Bees in Thailand, Dr. Bajaree Chuttong Department of Entomology and Plant Pathology, Faculty of Agriculture, Chiang Mai University and Dr. Hans Bänziger of Chiang Mai University for their hospitality during the author's time in Thailand.

We thank Dr H R Bhargava of Garden City University, affiliated with Bangalore University, Bangalore, Karnataka, India, for literary contributions to Meliponiculture in India.

Book cover design description:

Traditional Shan Tribal House with round-ended roof

The Shan State in Myanmar sits midway between South Asia and Indochina. An example of the Shan tribal house seems appropriate for the cover of this book. The Shan vernacular architecture accommodates all aspects of their surrounding nature. Built on stilts, there is sufficient room to keep domesticated animals below. Redrawn from Oranratmanee, R. (2018)

For the back cover, I feature a Tai Dam tribal house. They are part of the Tai people and ethnically similar to the Thai from Thailand, the Lao from Laos and the Shan from Shan State, Myanmar. The roof is similar to the Shan Tribal house, with an extension for the kitchen.

Inspired by a Tai Dam (Black Tai) Tribal house

The round-ended roof is adopted for the box hive because the round end provides extra protection for the hive nest entrance.

Contents

Sustainable Meliponiculture with Vernacular Architecture .... - 1 -

Volume 2 - South Asia & Indo-China .... - 1 -

Synopsis .... - 1 -

Acknowledgements .... - 2 -

Book cover design description .... - 3 -

Preface .... - 9 -

VOLUME 2, PART 1 .... - 10 -

~ SOUTH ASIA ~ .... - 10 -

The preamble to Part 1 .... - 11 -

Beekeeping in Pakistan .... - 11 -

Introduction to Part 1 .... - 14 -

Chapter 1 .... - 15 -

Diversity of Indo-Malayan Vernacular Architecture .... - 16 -

Chapter 2 .... - 17 -

Meliponiculture in India .... - 17 -

Stingless Bee Products and Their Economics .... - 18 -

    a) In the field of Agriculture: .... - 18 -

    b) Medicinal properties of stingless bee products: .... - 18 -

    c) Economics of stingless beekeeping .... - 18 -

Prospects of Meliponiculture .... - 19 -

South Indian Provinces .... - 20 -

Nicobar & Andaman Islands .... - 21 -

Eastern India Provinces .... - 22 -

Arunachal Pradesh .... - 25 -

Maharashtra, West India .... - 27 -

Madhya Pradesh .... - 28 -

Uttar Pradesh .... - 28 -

Chapter 3 .... - 29 -

Bengal and Bangladesh - Vernacular Roofing Styles .... - 29 -

Bengal Temple Architecture .... - 30 -

Chapter 4 .... - 34 -

Himalayan States Stingless Bees and Vernacular Structures .... - 34 -

State of Assam .... - 34 -

Meghalaya, NE India .... - 35 -

Sikkim, Himalayas, India .... - 36 -

Nagaland .... - 37 -

Meliponiculture in Nagaland .... - 39 -

Mizoram State in NE India .... - 39 -

Stingless Bees in Bhutan .... - 42 -

Chapter 5 .... - 43 -

Sri Lanka .... - 43 -

VOLUME 2, PART 2 .... - 47 -

~ INDO-CHINA ~ .... - 47 -

Preambles .... - 48 -

    People of Southeast Asia .... - 48 -

    Thailand - Chinese of Thailand .... - 48 -

    Malays, upland peoples, and new immigrants. .... - 49 -

Introduction to Part 2 .... - 51 -

Chapter 6 .... - 52 -

Indo-China Vernacular Architecture .... - 52 -

Myanmar Indigenous Architecture .... - 56 -

    Kayin Traditional Houses .... - 56 -

    Shan Tribal House .... - 57 -

Inle Lake, Nyaungshwe Township of Shan State .... - 58 -

Records of Meliponines on the Tenasserim Range (Rasmussen 2008) .... - 59 -

References for Vernacular architecture in Myanmar .... - 60 -

Vernacular Passive Design in Myanmar Housing .... - 61 -

Chapter 7 .... - 62 -

Stingless Beekeeping and Bee Plants in Thailand .... - 62 -

    Introduction .... - 62 -

    Stingless bee culture .... - 62 -

Differentiation of Stingless bee nests ..... - 63 -
Utilization of stingless bees ..... - 66 -
Stingless bee plants ..... - 67 -
Conclusion ..... - 68 -
Chapter 8 ..... - 69 -
Vernacular Architecture of Thailand ..... - 69 -
Traditional And Contemporary Houses in Thailand ..... - 70 -
Bamboo huts for the bee farm. ..... - 74 -
Wall cladding ..... - 75 -
Laos Traditional Housing ..... - 76 -
Chapter 9 ..... - 77 -
Kampuchea Vernacular Architecture ..... - 77 -
History ..... - 79 -
Houses and settlements ..... - 79 -
Chapter 10 ..... - 80 -
Types of Khmer House for Ordinary People ..... - 80 -
Pteas Pit ..... - 80 -
Pteas Rongdorl ..... - 80 -
Pteas Rongdeung ..... - 81 -
Pteas Kontaing ..... - 81 -
Pteas Khmer ..... - 82 -
Pteas Koeng ..... - 82 -
Traditional Rest House ..... - 83 -
Traditional rural Khmer house ..... - 83 -
Chapter 11 ..... - 85 -
Model Houses are redrawn, and historic text excerpts from https://www.salalodges.com/ ..... - 85 -
Traditional houses in a settlement ..... - 87 -
Traditional Waterside Khmer House ..... - 88 -
Chapter 12 ..... - 90 -
A Cham house in Ninh Thuan province, the Vietnam Museum of Ethnology ..... - 91 -
Hani Homes and Mushroom Houses ..... - 94 -

Meliponiculture in Vietnam	- 97 -
Meliponiculture	- 97 -
Model making for bee housing	- 99 -
References	- 100 -

Chapter 13 .... - 101 -

Perspectives on Meliponiculture in SE Asia	- 101 -
Introduction	- 101 -
What is meliponiculture?	- 101 -
What's in a Name?	- 101 -
Worker Development in *Tetragonula iridipennis* (stingless bees)*	- 102 -
The Queen is the most important bee in the colony	- 102 -
Purpose-made containers can simplify management and harvesting	- 102 -
Beekeeping in Laos and Cambodia	- 105 -
Stingless Bees In Phnom Penh	- 107 -
References	- 109 -

Chapter 14 .... - 110 -

Ancestor veneration among some Asian cultures .... - 110 -

Influences from China .... - 113 -

VOLUME 2, PART 3 .... - 116 -

~ Climate Adaptation & Fusion Architecture ~ .... - 116 -

Introduction to Part 3 .... - 117 -

Chapter 15 .... - 118 -

Changing Climate .... - 118 -

Thermal comfort in a tropical and sub-tropical climate .... - 118 -

Temperature and humidity recordings .... - 118 -

Thermal imaging .... - 118 -

Regional Heat Wave .... - 119 -

Chapter 16 .... - 123 -

Out of the box into the sphere .... - 123 -

Chapter 17 .... - 126 -

Fusion of Traditional architecture .... - 126 -

Sino-Portuguese and European Fusion .... - 128 -

Chapter 18 .... - 130 -

Diversity of Indo-Malayan Vernacular Architecture .... - 130 -

A touch of the 'Head Hunter' vernacular accent .... - 131 -

Chapter 19 .... - 136 -

Skull Collector Tribes .... - 136 -

Most recent headhunting news .... - 139 -

    Pak Ancah, Desa Penggilingan Padi, Sampit, Central Kalimantan .... - 139 -

    Traditional Betang House .... - 140 -

Chapter 20 .... - 141 -

Miniature Landscape and Meliponiculture .... - 141 -

Chapter 21 .... - 146 -

Meliponine Entomotourism .... - 146 -

    Introduction .... - 146 -

    Integrating Vernacular Architecture .... - 147 -

    Meliponini Tourism in the Jungles of Kalimantan .... - 150 -

    Stingless bee associations or co-operatives' collective show farms. .... - 150 -

Insect Museum in Brunei .... - 151 -

References for Entomotourism .... - 152 -

Appendix - Glossary of roof types .... - 153 -

List of Figures .... - 161 -

Index .... - 167 -

Bibliography .... - 172 -

Preface

This volume is set in three parts, although we cover a span from Pakistan to Vietnam. In Part 1, the scrutiny starts in the lower regions of Pakistan and then virtually moves into the West Indian Province of Maharashtra and down to South India, covering Karnataka and Kerala Provinces. In these southern regions, we gained much information from local beekeepers through social media communications and image exchange. From the lower tip of India, we move into the SE province of Tamil Nadu while exploring the type locality of some new species before hopping over to Sri Lanka. Still, on island cultural peculiarities, we cross the Bay of Bengal and next look at the Nicobar & Andaman Islands.

We leave the Islands and move towards Bangladesh, Assam and Eastern Myanmar. Here we are within the Himalayan states of NE India, where we interacted with some enthusiasts in Nagaland/ From there, with information from researchers in the Himalayan countries bordering North India, we will cover Bhutan and Nepal and appropriately complete the full circle.

For Part 2, we scrutinize the vernacular architecture of peoples in Myanmar and the Tenasserim Hills, which border Thailand. The diversity is immense as we cross the Thai border to Kampuchea and Vietnam. This area, formerly known as the French Indochina Peninsula, does not include Myanmar and Thailand in terms of architectural influence. The Champa architecture of Cham people here in Kampuchea and South Vietnam is unique and different from N. Vietnam, where Chinese architectural influence is mixed with some French accent roof ornaments. More variations abound in Laos, where accents of Malay origin are often intermingled with Khmer and Chinese influences and Thai Buddhist structures/. From Laos, we see hill tribes occupying lands of Laos, North Thailand and North Myanmar. Analysing roof pediments and rake and fascia board designs, we can observe similarities while having differences reflecting each region's architecture.

In Part 3, we look at climate change and thermal imaging of folk architecture. This part does not aim to be exhaustive in architectural elements but to assess the resultant designs that provide the ultimate thermal comfort within and ventilation efficacy. Overall, indigenous architecture has many basic Malay influences and accents. We glance at the fusion of traditional architecture with Sino-Portuguese and European Fusion. We have chapters on Miniature landscapes for Meliponiculture and also Entomotourism. Besides all these issues, ethnic and cultural backgrounds like ancestor veneration and headhunting history may reflect certain nuances in the indigenous dwellings and how they treat bees and the housing of their bees. Here, we cover a global aspect.

VOLUME 2, PART 1
~ SOUTH ASIA ~

The preamble to Part 1

Beekeeping in Pakistan

An estimated 10,000 beekeepers in Pakistan manage almost 600,000 *Apis mellifera* colonies and produce more than 12,000 tons of honey annually. Beekeeping in Pakistan is mainly focused in KPK[1] and central and northern regions of Punjab, but nowadays, it is growing rapidly[2].

The beekeeping community is at risk from the extreme weather conditions in Pakistan. This report by Dr. Khalid Ali Khan concerns mostly honeybees. However, studying the type localities of stingless bees in Northwest India and neighbouring South Pakistan, where the habitat is similar, shows a high probability of relevance for stingless bees.

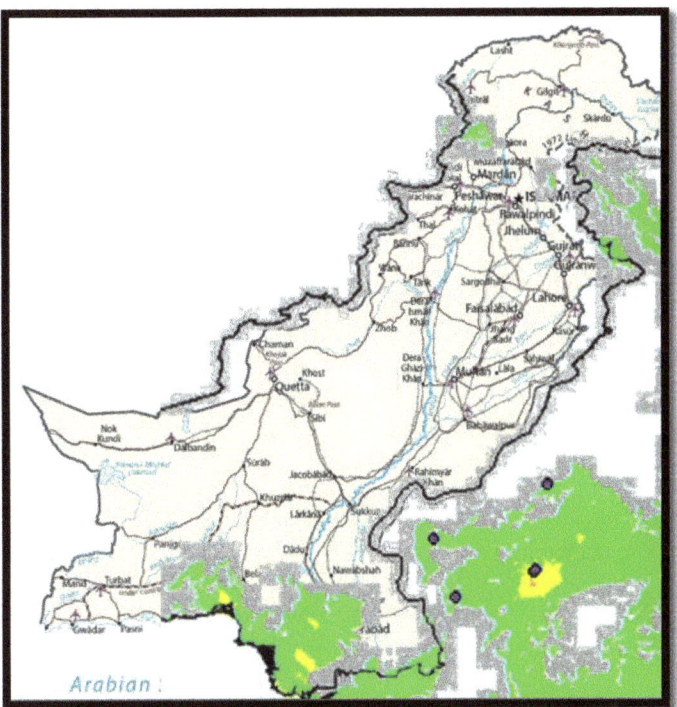

Figure 1 The superimposed green represents areas with the highest probability of habitat suitability and grey shows lower probabilities of habitat suitability. (Rasnusseb, 2013)

Since 95% of Pakistan's population are Muslims, we provide some ideas of stingless bee housing for beekeepers in that region who may be Meliponiculturists.

Working with straight beams for fast and easy construction allowed us to explore contemporary mosque designs. One may choose a model of the King Faisal Mosque in Islamabad, Pakistan, to incorporate the Bee Box Hive Roof Structure and direct sun protection. This model is something for Muslim Beekeepers anywhere in Indo-Malaya.

The roof design is a four-sided Butterfly roof. The finial on the roof summit and each minaret is an upturned crescent. This crescent makes for a perfect hook for the shade over the whole model. The roof can be wooden and covered with linoleum. The pediments at the gable ends can be perforated

---

[1] https://en.wikipedia.org/wiki/Khyber_Pakhtunkhwa
[2] Source: Beekeeping in Pakistan (History, Potential, And Current Status) By: Dr. Khalid Ali Khan Assistant Professor of Apiculture/Entomology at Unit of Bee Research and Honey Production, Faculty of Science, King Khalid University, Saudi Arabia. Posted: 22 July 2020. doi:10.20944/preprints202007. 0503.v1

with acrylic or Perspex sheets. The shade can be 80% black netting over a plastic sheet to protect from rain and direct sun.

Figure 2 Inspired by the King Faisal Mosque in Islamabad, Pakistan.

How is extreme weather changing Pakistan?

Besides the heat wave discussed above, on Tuesday, September 6, unprecedented flooding in Pakistan has left millions of people in ruin, with the country's central government and provincial authorities scrambling to provide urgent help to those who need it. "We have widened the earlier breach at Manchar to reduce the rising water level," provincial irrigation minister Jam Khan Shoro told the Reuters news agency. Already, 100,000 people have been displaced in efforts to keep the lake from overflowing, and if it breaches its banks, it could affect hundreds of thousands more, authorities said. (Source: https://www.aljazeera.com/news/2022/9/6/un-warns-of-humanitarian-crisis-in-flood-ravaged-pakistan)

Does Pakistan have earthquakes?

Pakistan is one of the most seismically active countries in the world, being crossed by several major faults. As a result, earthquakes in Pakistan occur often and are destructive. Pakistan geologically overlaps both the Eurasian and Indian tectonic plates. Sindh, Punjab and Azad Jammu & Kashmir provinces lie on the north-western edge of the

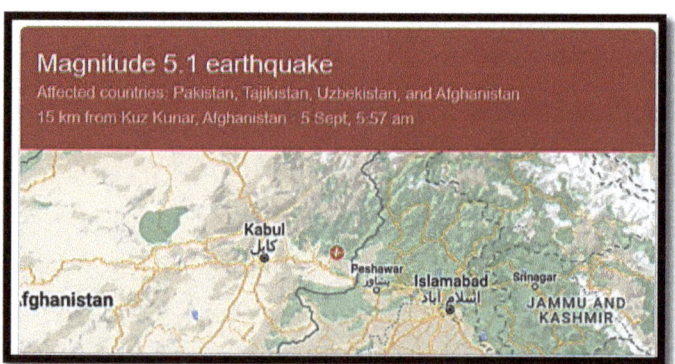

Figure 3 Latest earthquake in Pakistan 2022. Source: https://earthquake.usgs.gov/earthquakes/eventpage/us7000i57m/executive

Indian plate in South Asia. Hence, this region is prone to violent earthquakes as the two tectonic plates collide. Source: https://en.wikipedia.org/wiki/List_of_earthquakes_in_Pakistan

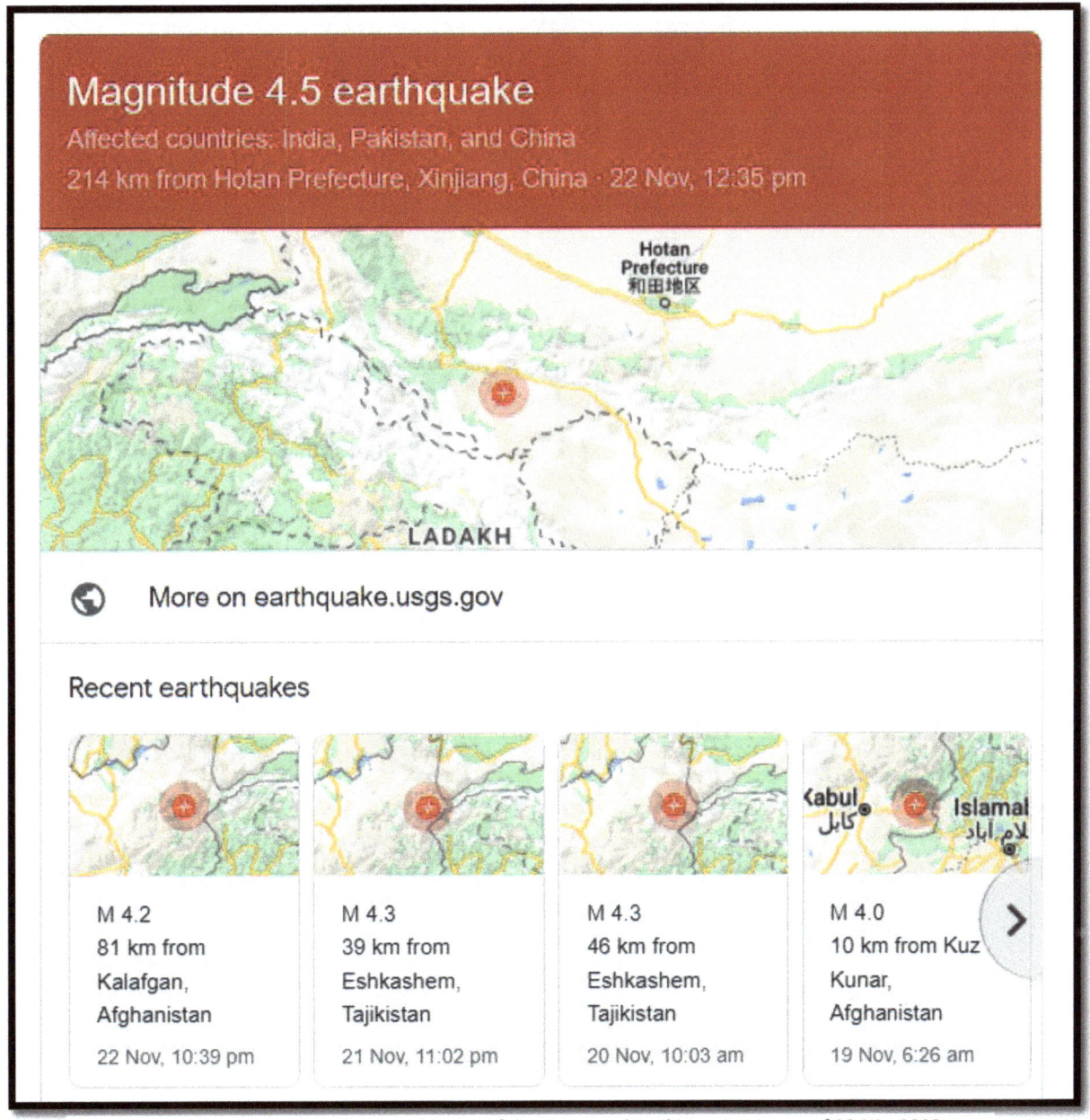

Figure 4 Recent Earthquakes in Pakistan, Afghanistan and Tajikistan Region as of 22 Nlv, 2022

Nearby Meliponiculture activities are in the neighbouring Punjab region and west Nepal, as shown in the charts below. Anything north does not have a climate suitable for stingless bee-keeping.

Introduction to Part 1

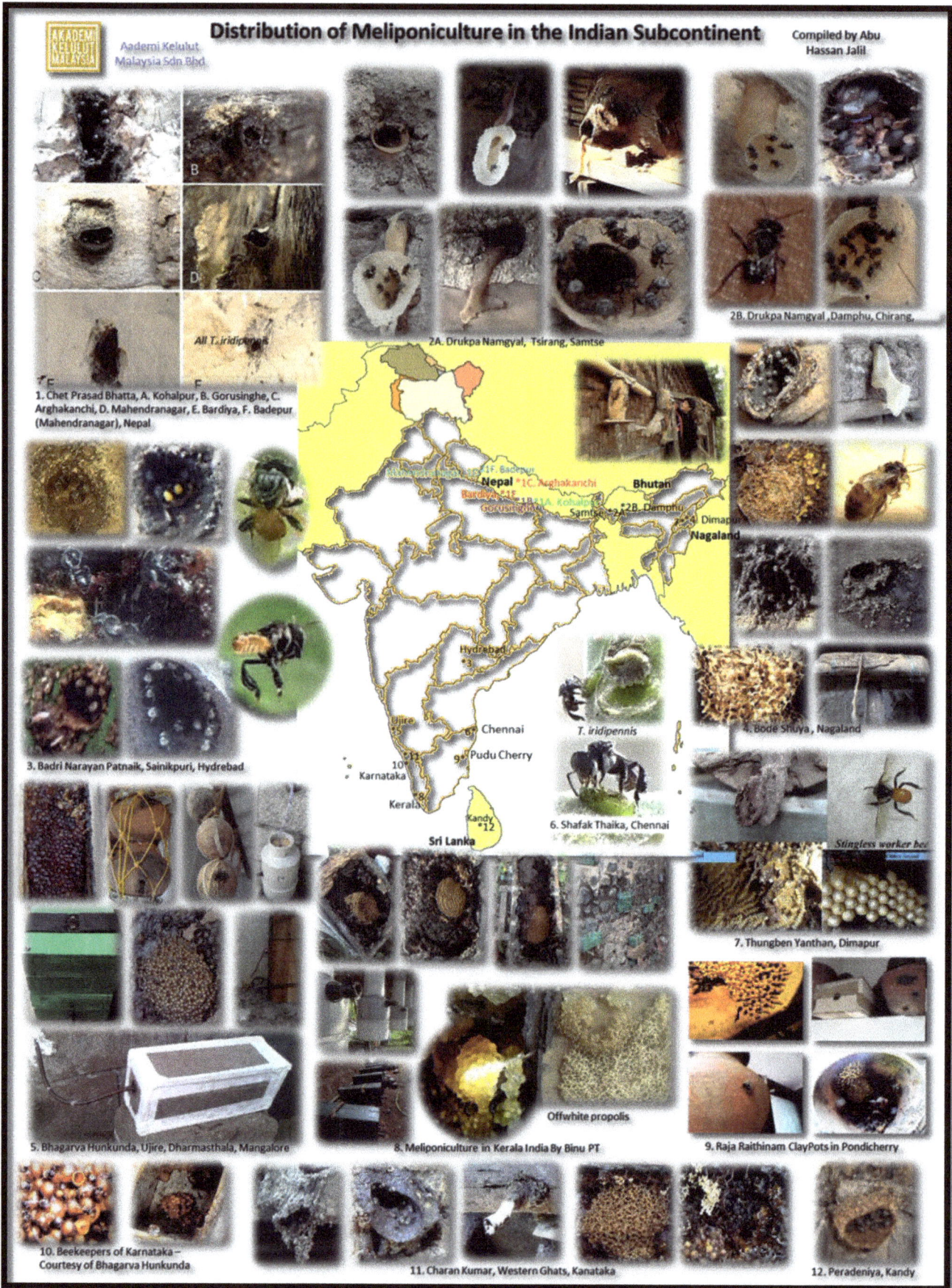

Figure 5 Distribution of Meliponiculture activities in the Indian Subcontinent

Chapter 1

Indian Subcontinent

Type localities for the species of stingless bees from the Indian subcontinent. a. *Tetragonula iridipennis*, b. *Lisotrigona mohandasi*, c. *L. cacciae*, d. *T. ruficornis*, e. *T. bengalensis*, f. *Lepidotrigona arcifera*. *T. praeterita* was described from "Ceylon"- no further indication of the locality.

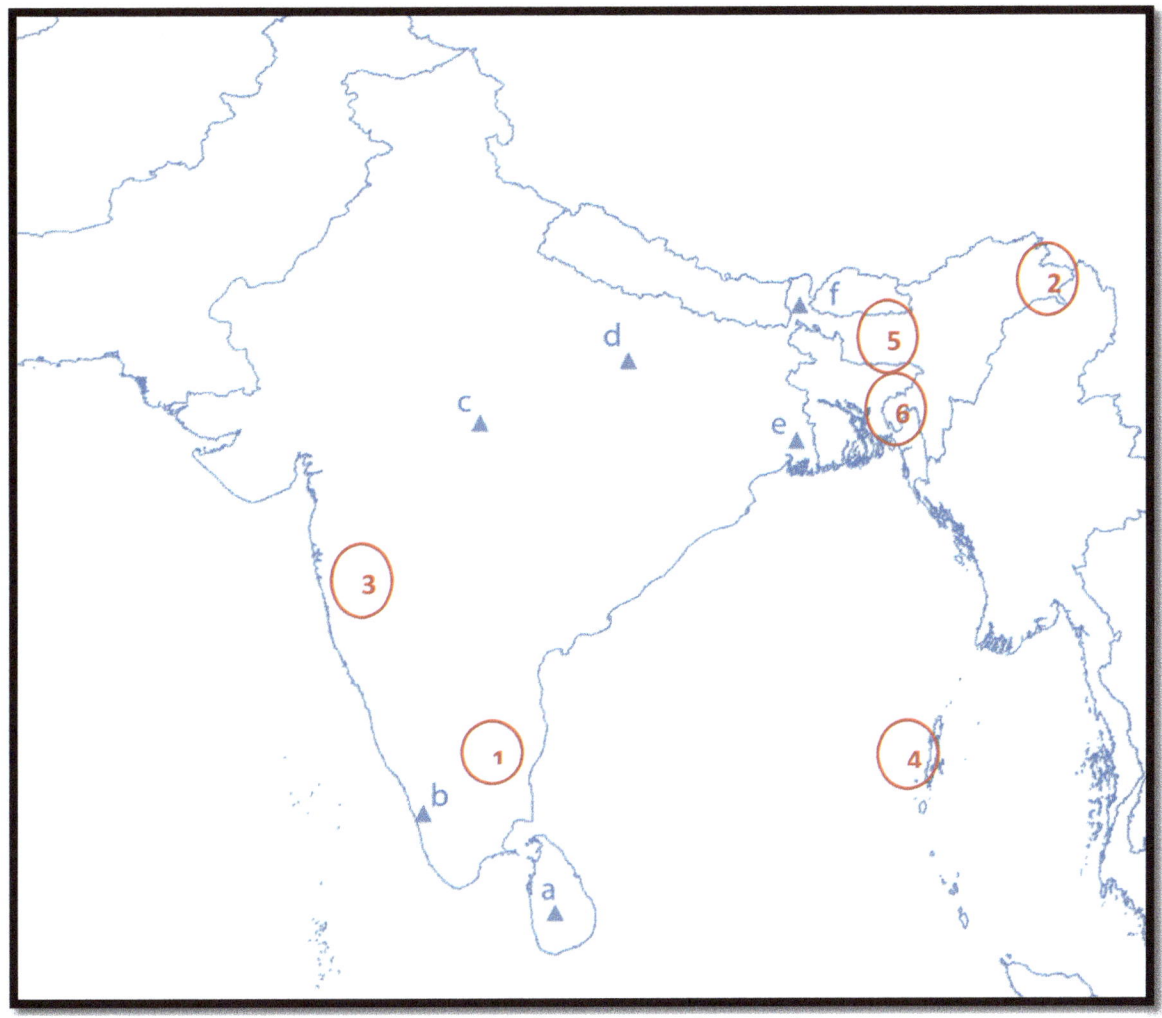

Figure 6 Type localities for the species of stingless bees

Indications of new species up to 2015: 1. *T. (Flavotetragonula) calophyllae* Shanas and Faseeh, n. sp.; *T. (T.) perlucipinnae* Faseeh and Shanas, n. sp.; *T. (T.) travancorica* Shanas and Faseeh, n. sp.;. *Lisotrigona chandrai* Viraktamath and Sajan Jose 2017,; 2. *T.(T.) gressitti* (Sakagami 1978)[3]; 3. *L. revanai* Viraktamath and Sajan Jose 2017; 4. *T.* aff, *laeviceps* (Smith 1857)[4]; 5. *T. kyrdemkulaiensis* Viraktamath and Rojeet sp. n.; 6. *T. srikantanathi* Viraktamath sp. n.

---

[3] Type locality of *T. gressitti* is Vietnam. Rathor and Rasmussen have reported this species from India.
[4] Type locality of *T. laeviceps* is Singapore (Please see Rasmussen 2013). However, bees similar to *laeviceps* have been reported from Bengaluru (=Bangalore) and Andaman and Nicobar Islands by Sakagami (1978). Recent collections by Dr. Viraktamath (personal communication) have revealed that stingless bees occurring in Andaman and Nicobar Islands are *Tetragonula* aff *malaipanae*.

Diversity of Indo-Malayan Vernacular Architecture

Figure 7 Vernacular Architecture in the Indian Subcontinent

Architectural diversity is enormous[5], from Nicobar and the Andaman Islands Straw Huts to traditional Wada architecture in Maharashtra, from old houses in Bengaluru to the Chala roof styles in Bengal and other Himalayan states. The materials used are what is locally available in abundance.

Sun-dried mud bricks with wood frames, straw-thatched roofs, and most build-ups with folk labour and expertise are void of professional architects. The building techniques practised by generations of untrained in formal architectural design constitute handed down indigenous
knowledge. These practices have been proven stable in past calamities and devastation by the extremes of natural conditions and situations. These techniques are the essence[6] of vernacular architecture.

Chapter 2

Meliponiculture in India

By: Dr H. R. Bhargava (http://scholar.google.co.in/citations?user=2l_iwPoAAAAJ&hl=en)

Edited by Dr. Shashidhar Viraktamath (Emeritus Scientist, UAS, GKVK, Bengaluru) and Dr. David Roubik.

Figure 8 A Salvaged South Indian House in the Heritage Village in Manipal, Karnataka. Drawn by AHJ.

Stingless bees are called Dammar bees or Dammer bees (Damar is a resin of Dipterocarp trees). Locally stingless bees are called by different names like *putka* in Sikkim and Nepal, *ngapsiwor*, *ngaphamang* and *ngapkhyndew* in the Khasi language, *Cherthneecha* and *arakki* in Kerala, and *Misri Jenu* in Karnataka.

Stingless bees are widely distributed worldwide in various parts of India-Burma, Malayan and Australian regions. There is a diversification in their species level. Here is the list of the stingless bee species distributed in different regions of India.

a) Northeastern regions.

i) *Lepidotrigona arcifera* (Cockerell 1929)

ii) *Tetragonula iridipennis* (Smith, 1854)

iii) *Tetragonula bengalensis* (Cameron, 1897)

b) **Southern Region**

i) *Tetragonula ruficornis* (Smith 1870)

c) Andaman and Nicobar Islands (a union territory of India)

i) *Tetragonula aff malaipanae.* (See also p. 19)

---

[5] https://pegasusorg.com/f/vernacular-architecture---the-past-the-present-and-the-future
[6] https://en.wikipedia.org/wiki/Vernacular_architecture

d) Neighbouring island nation of Sri Lanka

i) *Tetragonula praeterita* (Walker 1860)

Stingless Bee Products and Their Economics

The importance of stingless bees and their products has placed them in a unique position.

a) In the field of Agriculture:

Stingless bees pollinate various crops and enhance crop production in agriculture, like tomatoes and other pollination-dependent crops. It is quite advantageous to rear or keep these bees on their farms to enhance crop productivity through pollination, and these bees are harmless compared to the *Apis* species (stinging bees).

b) Medicinal properties of stingless bee products:

i) **Honey**

The stingless bees' honey is a rich source of antioxidants and flavonoids. The honey of stingless bee species has rich amounts of flavonoids and flavanols with a high source of nitric oxide scavenging activity.

ii) **Propolis**

The other important product of stingless bees having high pharmacological value is Propolis. Propolis is a resinous material consisting of bee saliva and beeswax with exudates from plant buds, tree bark injury, or other botanical sources[7]. Propolis has a wide variety of antibacterial and antiviral properties. The medicinal efficacy of Propolis depends upon its composition and various factors like bee visitation or preference during foraging and the amount of pollution present in the environment.

c) Economics of stingless beekeeping

Stingless bee products have a good price in the market due to their pharmacological properties. As the amount of honey obtained from the colonies is less than a kilogram per colony, depending on the size of the colony generates a great demand for the product. This, in turn, allows the farmers to raise the bees (beekeeping). In this way, there was also a great demand for the colonies, which provided an opportunity to raise the colonies for marketing. Per kilogram of stingless bee honey costs around Rupees1,500 to 2,000 INR in a local market. With appropriate and innovative processing technologies, the honey will have a high commercial value in the International market.

---

[7] https://arboretum.harvard.edu/stories/plant-exudates-and-amber-their-origin-and-uses/#:~:text=What%20are%20plant%20exudates%3F,or%20some%20other%20plant%20pathology.

The Propolis of stingless bees is also as valuable as that of honey. Propolis is processed, extracted scientifically, and supplied to the pharmacological industries to prepare drugs. Propolis also has a high commercial value similar to stingless bees' honey. The pollen is also scraped out from the pollen pots of the stingless bee colony and sold in the market due to its high proteinaceous content.

Prospects of Meliponiculture

The Honey industry in India can become a major foreign exchange earner if International standards are met. Beekeeping, including Meliponiulture, is an age-old tradition in India. Honey production is a lucrative business, and it generates employment. The informal sector provides up to 70% of honey, including stingless bee honey and bees wax market in India. Indian stingless bee honey offers tremendous export potential because of the diversity in flora. There is a need to chalk out a suitable export strategy to trap its potential.

Kerala state is rich in diversified flora, including rubber plantations that provide a good source of nectar for the bees, suitable for rearing stingless bees. In Kerala, the University of Kerala proposed a project named "Oru Veettilum Oru Cherutheneecha koodu" (One stingless bee colony in every house) to provide each household with a colony of stingless bees. This has become an example for other State Universities for the sustainability and enhancement of stingless beekeeping.

South Indian Provinces

Kerala (Photos courtesy of Binu PT)

Figure 9 a) & b) Madhusree Bee Farm c) Stingless bee trapping d) & e) PVC pipe housing for bees f) Painted wooden box hives

Tamil Nadu.

Flavotetragonula, Etym: The subgeneric name is derived from flavus in Latin for yellow.

Abstract : A new subgenus of stingless bees, *Flavotetragonula* Shanas, subgen. n. is established, and three new species, *Tetragonula (Flavotetragonula) calophyllae* Shanas and Faseeh, n. sp., *Tetragonula (Tetragonula) perlucipinnae* Faseeh and Shanas, n. sp. and *Tetragonula (Tetragonula) travancorica* Shanas and Faseeh, n. sp. is described from Southern India, based on workers.

calophyllae - ***Tetragonula (Flavotetragonula) calophyllae*** Shanas and Faseeh, n. sp. (1), Etym: The specific epithet is after the generic name of the endangered tree Calophyllum inophyllum, on which the first feral colony was observed.

perlucipinnae - ***Tetragonula (Tetragonula) perlucipinna****e* Faseeh and Shanas, n. sp., Etym: The specific epithet, based on the Latin words perlucidulus and pinna, alludes to the transparent wing.

travancorica -***Tetragonula (Tetragonula) travancorica*** Shanas and Faseeh, n. sp., Etym: This species is named after the erstwhile kingdom of Travancore.

Remarks: The Kingdom of Travancore (/ˈtrævənkɔːr/) (Thiruvithamkoor) was an Indian kingdom from c. 870 until 1949. The Travancore Royal Family ruled it from Padmanabhapuram and later Thiruvananthapuram. At its zenith, the kingdom covered most modern-day central and southern Kerala with the Thachudaya Kaimal's enclave of Irinjalakuda Koodalmanikyam temple neighbouring Kingdom of Cochin, as well as the district of Kanyakumari, now in the Indian state of Tamil Nadu.

Lisotrigona chandrai Viraktamath and Sajan Jose 2017(2) [Etym: Dedicated to Indian taxonomist Dr. Chandrashekhar A. Viraktamath]

Remarks: This is named after a well-known Indian Taxonomist, Dr. Chandrashekhar A. Viraktamath, popularly known as "Chandra" among his international colleagues and friends. Dr. CAV (that is how he is known among his Indian students and colleagues) has completely dedicated his life to the study of Indian insects, especially leafhoppers. He is a mentor and guides a host of Indian and International students, including the authors of this paper. This new species of stingless bees is named as a tribute to this great Indian personality. (see **Note)

Nicobar & Andaman Islands

The Nicobarese people are an Austroasiatic-speaking people of the Nicobar Islands, a chain of islands in the Bay of Bengal north of Sumatra, forming Part of the union territory of Andaman and Nicobar Islands, India. Only 12 of the 19 islands are inhabited. The largest and main island is Great Nicobar. The term Nicobarese refers to the dominant tribes of the Nicobar Islands. On each island, the people have specific names, but together they are the Nicobarese. They call themselves Holchu, which means "friend."

Figure 10 Andaman and Nicobar Islands village hut

The villages on the islands consist of sporadically placed huts strewn about in designated areas. The huts are normally round with dome-shaped roofs. They are typically raised above the ground and have ladders that the residents pull up after they climb into the huts at night. Source:[8]

Sakagami (1978) examined a worker of presumably *T.* aff. *laeviceps* from the **Andaman** Islands. Recent collections by Dr. Viraktamath (personal communication) have revealed that stingless bees occurring in Andaman and Nicobar Islands are *Tetragonula aff malaipanae*.

---

[8] https://en.wikipedia.org/wiki/Nicobarese_people#/media/File:Crafts_Museum_New_Delhi_3_Sep_2010-11.JPG

Eastern India Provinces

"State Tribal Fair- 2020", Bhubaneswar, Odisha, India:

1. Munda House

Figure 11 Munda_House_at_State_Tribal_Fair-020__Bhubaneswar

The Munda are mainly concentrated in the south and East Chhotanagpur Plateau region of Jharkhand, Odisha and West Bengal. The Munda also reside in adjacent areas of Madhya Pradesh and portions of Bangladesh and Tripura. They are one of India's largest tribes. Munda people in Tripura are also known as Mura. Source[9]

2. A tribal house of the Kisan community from Odisha

Figure 12 Kisan tribal house at the 2020 Odisha State Tribal Fair, Bhubaneswar

The Kisan or Nagesia are a tribal group in Odisha, West Bengal and Jharkhand. They are traditional farmers and food-gathering people.

Other populations live in western West Bengal's Malda district and western Jharkhand's Latehar and Gumla districts. Source[10]

3. Oraon House

Figure 13 Oraon House at 'State Tribal Fair-2020' Bhubaneswar, India

The Kurukh or Oraon or Dhangar (Kurukh: Kuṛukh and Oṛāōn), also spelt Uraon or Oraon, are a Dravidian ethnolinguistic group inhabiting the Indian states of Jharkhand, West Bengal, Odisha and Chhattisgarh. In Maharashtra, Oraon people are also known as Dhangad or Dhangar. Source[11]

---

[9] https://en.wikipedia.org/wiki/Munda_people
[10] A typical house of tribals of the Kisan community from Odisha
[11] https://en.wikipedia.org/wiki/Kurukh_people

4. Santal House

Figure 14 Santal house at 2020 Odisha Tribal Fair, Bhubaneswar

The Santal or Santhal are the largest tribe in the Jharkhand and West Bengal states of India in terms of population and are also found in the states of Odisha, Bihar and Assam. They are the largest ethnic minority in northern Bangladesh's Rajshahi and Rangpur Division. Source[12]

5. Gadaba Hut

The **Gadaba** or Gutob people are an ethnic group in eastern India. They are designated Scheduled Tribe in Andhra Pradesh and Odisha. Their socioeconomic life is based on farming and daily labour.

Figure 15 Left: A Gadaba hut, Koraput, Odisha; Middle: A Village complex in Andhra Pradesh; Roght: A typical house of tribals of the Gadaba community from Odisha.

They are involved in both Slash-and-burn and plough cultivation. They live in permanent villages. Since the early 1980s, the Gadabas have largely been displaced from their villages by the building of hydroelectric dams and the resulting lakes. Source[13]:

6. Sora House

Figure 16 Lanjia Sora house at Odisha State Tribal Fair, Bhubaneswar

The **Sora** (alternative names and spellings include Saora, Saura, Savara and Sabara) are a Munda ethnic group from eastern India. They live in southern Odisha and north coastal **Andhra Pradesh**. They are animists and believe in many deities and ancestral spirits. The Soras' religion is very elaborate and deep-rooted.

---

[12] https://en.wikipedia.org/wiki/Santal_peopleg

The Sora family is polygamous. They practice shifting cultivation, and the men hunt. A weekly market, *shandies*, plays an important role in the economy and cultural exchanges with other tribes and Western cultures. Source[14]:

7. Bathudi House

Figure 17 A typical house of tribals of the Bathudi community from Odisha.

The Bathudi are a community found mainly in the northwestern part of Odisha. Their houses are mainly made of mud walls and thatched roofs. Typical household holds goods like stringed Charpoys; aluminium, bell metal and earthen utensils; bow and arrow; fishing tools, mats, etc. Source[15]

8. Khond house.

Khonds (also spelt Kondha, Kandha, etc.) are an indigenous Adivasi tribal community in India. Traditionally, hunter-gatherers are divided into the hill-dwelling Khonds and plain-dwelling Khonds for census purposes.

They are a designated Scheduled Tribe in Andhra Pradesh, Bihar, **Chhattisgarh**, Madhya Pradesh, Maharashtra, Odisha, Jharkhand and West Bengal. Source[16]

Figure 18 A traditional Khond house.

Recent findings of *Tetragonula* spp. nov. 2022 in India

Tetragonula vikrami Viraktamath sp. n. **Etym**.: This species is named as a tribute to Mr. Vikram, a Covid-19 warrior, later became a victim of it. **Holotype**: Male: **Karnataka**: Mankalale. 25. iii. 2019. Coll. Bhat S. deposited at UASB. **Paratypes**: 44 males and 24 females with the same collection data deposited at UASB. One paratype male will be deposited each at IARID and ZSIK.

Tetragonula sumae Viraktamath sp. n. **Etym**.: This species is named in honour of Mrs. Suma Viraktamath, a constant source of inspiration, encouragement and support to the senior author. **Holotype**: Male: **Tamil Nadu**: Salem. 15. vi. 1915, Coll. Dutt GR deposited at UASB. **Paratypes**: Three males and two females with the same collection data deposited at UASB.

---

[14] https://en.wikipedia.org/wiki/Sora_people
[15] https://en.wikipedia.org/wiki/Bathudi_Tribe
[16] https://en.wikipedia.org/wiki/Khonds

Tetragonula ashishi Viraktamath and Jagruti sp. n. **Etym**.: This species is named in honour of Dr. Ashish Kumar Jha, who collected and spared these bees for our studies. **Holotype**: Male: **Maharashtra**: Nagpur, 26. ix. 2019, Coll. Paratypes: 21 males and 50 females with the same collection data but Coll. Ashish Kumar J. deposited at UASB. Jagruti Roy will be deposited at UASB; one male and one female will be deposited each at IARID and ZSIK.

Tetragonula shishirae Viraktamath sp. n. **Etym**.: This species is named after Mrs. Shishira, who collected these bees and spared them for our studies. **Holotype**: Male: **Rajasthan**: Udaipur, 11. iii. 2019, Coll. Shishira D. deposited at UASB. **Paratypes**: 02 males and 73 females with the same collection data deposited at UASB; one female paratype will be deposited each at IARID and ZSIK.

Tetragonula shubhami Viraktamath sp. n. **Etym**.: This species is named after Mr. Shubham Rao, who collected and spared these bees for our studies. **Holotype**: Male: **Chhattisgarh**: Bardebhata, 06. xi. 2019, Coll. Shubham Rao deposited at UASB. **Paratypes**: Four males, **Chhattisgarh**: Pushpal, 15. iv. 2020, Coll. Shubham Rao, one male, Chhattisgarh: Godre, 5. ix. 2019, Coll. Shubham Rao deposited at UASB. Fifty-nine females (10 mounted 49 in 95% alcohol vial), **Chhattisgarh**: Bardebhata, 06. xi. 2019, Coll. Shubham Rao deposited at UASB. One female paratype will be deposited each at IARID and ZSIK.

Arunachal Pradesh

The **Adi** people are one of the most populous groups of indigenous peoples in the Indian state of Arunachal Pradesh. The majority of Adi traditionally follow the tribal Donyi-Polo religion. Worship of gods and goddesses like Kine Nane, Doying Bote, Gumin Soyin Pedong Nane, etc., and religious observances are led by a shaman called Miri (who can be a female). Each deity is associated with certain tasks and acts as a protector and guardian of various topics related to nature, which revolves around their daily life. This includes the food crops, home, rain, etc. Source[17]:

Figure 19 A traditional Adi hut (interestingly, it looks similar to the Ibaloi House in The Philippines)

---

[17] https://en.wikipedia.org/wiki/Adi_people

The Jingpo people are an ethnic group who are the largest subset of the Kachin peoples, which largely inhabit the Kachin Hills in northern Myanmar's Kachin State and neighbouring Dehong Dai and Jingpo Autonomous Prefecture of China. There is also a significant Jingpo community in northeastern India's Arunachal

Figure 20 Singpho dwelling in Arunachal Pradesh

Pradesh, Assam, and Taiwan. While they mostly live in Myanmar, the Kachin are called the Jingpo in China and Singpho in India – the terms are considered synonymous. They live in Arunachal Pradesh in the Lohit and Changlang, and Assam inhabits the district of Tinsukia and is scattered in some other districts like Sivasagar, Jorhat and Golaghat. Singpho dwellings are usually two stories built out of wood and bamboo. The houses are oval; the first floor serves as a storage and stable, while the second is utilized for living quarters. Source[18]

Bee species reported in **Arunachal Pradesh**: Tetragonula gressitti (Sakagami 1978)

Tetragonula (Tetragonula) gressitti Sakagami 1978: 214–216: Holotype (not located, see Rasmussen (2008), worker): Type locality: VIETNAM, Lâm Đồng province in the Central highlands.

This species is characterized by distinctly melanic colouration, including the entire antenna's ventral side and relatively long malar space and scape. This species was recently reported from Hunli and Pashighat, Lower Dibang Valley district of **Arunachal Pradesh** in India's extreme northeastern Himalayan region, close to China (Rathor et al.2013).

Figure 21 Rendering of T. gressitti (Sakagami 1978) by AHJ

---

[18] https://en.wikipedia.org/wiki/Jingpo_people

Maharashtra, West India

The Katkari, also called Kathodi, are an Indian tribe from **Maharashtra**. The Katkari were once a forest living in the Western Ghats of Maharashtra, with a special relationship to forest creatures.

While no longer a forest people, Katkari's knowledge of forest resources remains with them. Katkari, living close to forested areas, still consumes over 60 different uncultivated plants and over 75 different animals and birds, gathering these with incredible ingenuity and skill. Source[19]

Figure 22 Katkari, also called Kathodi dwelling
http://www.rainforestinfo.org.au/projects/india/Katkari.htm

Figure 23 This typical home belongs to the Tadvi Bhils in Maharashtras Satpuda region

This typical home belongs to the Tadvi Bhils[20] in **Maharashtra's** Satpuda region. The Tadvi Bhil is a tribal community in India in Maharashtra, Gujarat, Madhya Pradesh and Rajasthan. This community is a resident of Gujarat, Madhya Pradesh, and Maharashtra's Satpuda Hills spread.

Sawantwadi is located at 16°N 73.75°E in the Sindhudurg district of **Maharashtra**. It has an average elevation of 22 metres (72 ft) above mean sea level. It is the administrative headquarters of the Sindhudurg district. Sawantwadi is situated on the west coast of Maharashtra, India, and is bounded by the Arabian Sea to its west and the Western Ghats to its east.

Figure 24 Sawantwadi dwelling

Bee species recorded in the locality:

Lisotrigona revanai Viraktamath and Sajan Jose 2017[(2)] [Etym: The species is named after a famous Hindu God, "Revanasiddeshwara". Type Loc. It is collected in **Sawantawadi**, **Maharashtra**, India. (Pers. Comm. Viraktamah, 2017)]

Remarks: The species is named after a famous Hindu God, "Revanasiddeshwara". The temple is located near the type locality of the new species on a hillock known as "Revanadi" in Maharashtra. Also known as a **Theonym** - A **theonym** (from classical Greek *theos* "god" and *-onym* "name") is a proper name of a deity.

---

[19] https://en.wikipedia.org/wiki/Katkari_people#/media/File:Katkari_village_enclosed.tif
[20] https://commons.wikimedia.org/wiki/File:Residence_of_Tadvis_in_Village.jpg

Madhya Pradesh

At Mardhari village in the buffer zone of Bandhavgarh National Park, vegetables like pumpkins are grown on the roofs of homes and cattle sheds. A timber framework supports a mud structure at Sarekha village in Kanha.

Figure 25 Mardhari village vegetables like pumpkin are grown on the roofs of homes

Bee species relevant to Madhya Pradesh locality:

Lisotrigona cacciae (Nurse 1907)
Melipona cacciæ Nurse 1907: 619: Lectotype (BMNH 17b.1103, worker): examined, "LECTOTYPE" (blue border), "Type" (red border), "B.M. TYPE / HYM. / 17B.1103", "Melipona / cacciæ / (Nurse)", "Col. C.G. Nurse / Collection. / 1920-72", "Hoshung- / -abad / Type", "LISOTRIGONA / cacciae (Nurse) / det. M. S. Engel, 1999", "LECTOTYPE / Melipona / cacciae Nurse / design. J. S. Moure, 1961 / (ref: Engel, Oriental Insects, 2000)". **Type locality: INDIA, Madhya Pradesh**, Hoshangabad [ca. 23.25°N, 78.2°E].

Uttar Pradesh

Traditional Boatmen and Fishermen Tribes

The **Dhimar** are a caste in India, sometimes called a subcaste of boatmen. Source[21]

Figure 26 "Dhīmar or fisherman's hut." from The Tribes and Castes of the Central Provinces of India Volume II Author: R. V. Russell

The **Mallaah** are the traditional boatmen and fishermen tribes or communities of North India, East India, Northeastern India and Pakistan. A significant number of Mallaah are also found in Nepal and Bangladesh. In the Indian state of Bihar, the term Nishad includes the Mallaah and refers to communities whose traditional occupation centred on rivers. Source[22]

Bee species relevant to Uttar Pradesh locality:

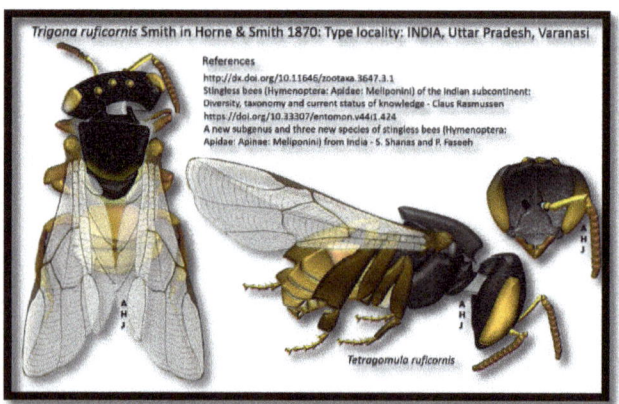

Figure 27 Figure 18 Rendering of T. ruficornis (Smith in Horne & Smith 1870) by AHJ

Tetragonula ruficornis (Smith in Horne & Smith 1870)
Trigona ruficornis Smith in Horne & Smith 1870: 185, 194: Lectotype (BMNH, worker): examined, "India" (typed, with handwritten reverse "69 / 86"), "SYNTYPE" (blue border), "SYNTYPE worker symbol] / Trigona / ruficornis / F. Smith, 1870: 194 / det. D. Notton, 2012"). In addition, "LECTOTYPE *Trigona ruficornis* Smith Design. C. Rasmussen 2013"; Type locality: INDIA, Uttar Pradesh, Varanasi (formerly Benares) (on April 4th, 1863) [25.28°N, 82.96°E].

---

[21] https://en.wikipedia.org/wiki/Dhimar#/media/File:Dhimar_caste_hut.jpg
[22] https://en.wikipedia.org/wiki/Mallaah

Chapter 3

Bengal and Bangladesh - Vernacular Roofing Styles

Bengal is a land of culture and art carved by innumerable temples. These temples are vernacular architectural adaptations of the traditional Bengali hut that portrays a typical Chala roof. The Chala roof is a gable type with two-, four- or eight–sloping roofs with curved edges or cornices meeting at a curved ridge. The slope of the roof performs the drainage function against rainfalls. The curved

Figure 28 Vernacular Architecture of West Bengal- Zonal Geography Classification According to Climatic Conditions

structure is due to the flexibility of roofing material, i.e., bamboo and thatch used in Bengali huts. The interior curvature of a traditional hut roof supported by bamboo or wooden posts forms a dome. To increase longevity in high rainfall areas, temples comprised regionally available bricks and terracotta. Keshta Raya temple (Bishnupur), Raghavesvara temple (Diknagar), and Siva temple (Amadpur) are some of the preserved Bangla temples. Source[23]:

Even today, the rural huts are built of concrete and bricks with *do-* or *cahu-chala* roofs made of corrugated iron sheeting or clay tiles, sustaining a legacy of Bengal's traditional vernacular architecture.

Bengal Temple Architecture most commonly resembles those old-style mud-plastered thatched huts with either 1, 2, 3, 4, or 8 distinct layers of slanting rounded roofs perfectly dotting the region's landscape. The temples built between the 16th and 19th centuries in this region resemble one of the most distinctive groups of brick monuments in India. Due to the multiple artistic influences on the

---

[23] http://www.authorstream.com/

region, Bengal temple architecture showcases various construction techniques and forms, mainly in their curved roofs and arched entrances during this period. In addition to this, they also reflect a fusion of local Bengal expressions and Islamic architecture.

Bengal Temple Architecture

1. *Do-chala*

The *do-Chala*, also known as *Ek-Bangla*, is a structure that has two sloping roofs with curved cornices that meet at curved ridges. In terms of the internal structure, there is a rectangular chamber enclosed under a vaulted roof. This particular style reflects the single-celled huts of the State and has been seen even in Islamic Architecture. Source[24]:

Figure 29 Do-chala roof- Left: Damodar temple of Siuri in Birbhum district; Right: Jor Bangla Temple, Bishnupur with a curved Do-chala style roof

The second temple is in Bishnupur. Source[25]:

2. *Char-chala*

The Char-Chala temples have four rectangular roofs meeting at one point. The edges of the chala, along with the cornices, are carved. It is quite a rare roofing style as far as the temples of Bengal are concerned, and you will only find a few structures with this roofing style in Nadia, Murshidabad, and Birbhum districts. Source[26]: The second structure is the Palace in Deeg, Rajasthan. Source[27]:

Figure 30 Char-chala style roof; Left: Palpara Temple – Nadia 2011; Right: Palace in Deeg, Rajasthan (in 1750)

Figure 31 At-chala roof – Atchala and Pancha Shiva Mandir in Pathra; Right: Antpur Radhagovindjiu Temple;

3. *At-chala*

The *At-Chala* can be best described as a variation of the *Char-chala* temple. Think of a *Char-chala* temple with a truncated roof and another *Char-chala* temple added on top of it, and that is what the roof of an

---

[24] https://commons.wikimedia.org/wiki/File:Damodar_temple_of_Siuri_in_Birbhum_district_of_West_Bengal_01.jpg
[25] https://en.wikipedia.org/wiki/Architecture_of_Bengal#/media/File:Jor-Bangla_Temple.jpg
[26] https://upload.wikimedia.org/wikipedia/commons/3/31/Palpara_Temple_-_Nadia_2011-10-05_050412.JPG
[27] https://en.wikipedia.org/wiki/Bengal_roofs#/media/File:Deeg_Palace.jpg

At-Chala temple looks like. This roofing style is widely noted in the Hugli, Medinipur, Howrah, and Bankura districts. The Malancha Dakshina Kali temple in Medinipur is the finest example of the At-Chala roofing style[28]. The second structure is Antpur Radhagovindjiu Temple[29]:

4. *Ratna*

The *Ratna* design shows a marked deviation from the sloping or **chala** roofing styles. These temple roofs are flat and surmounted by pinnacles, known as *ratnas* or *churas*. The origin of this style is unclear because there are Islamic and Hindu precedents of structures with one turret or more than that[30]. The second temple is in the Hooghly district[31]:

Figure 32 Ratna roof – Left: Ram Chandra Temple, Guptipara, Hooghly district; Right: Radha Govinda Temple Bishnupur Bankura

5. *Ek-ratna*

Figure 33 Ek-ratna roof – Ekratna Temple of Gopinath, Khard Radhakantapur

The *Ek-Ratna*, or single-towered structure, is a simplified version of the *ratna* style. This temple was particularly favoured by the Malla rulers, who had built many such structures at the seat of their power in Bishnupur. It is also important to mention that most of these temples, including the pinnacles and the cornices, are made of laterite, not brick.

Source[32]: The second temple is in Bansberia City, Hooghly district[33]:

6. *Dalan*

These flat-roofed styles of the temple became popular during the 19th century, especially in the district of Medinipur. In fact, in

Figure 34 Dalan roof – Left: Dalan Temple inside Kaviraj Bari at Mankar situated in Purba Bardhaman district; Right: A 17th century haveli (a mansion) in Old Dhaka

Medinipur, there is a clear distinction between the large flat roofs, known as *dalan*, and the smaller

---

[28] https://commons.wikimedia.org/wiki/File:Atchala_and_pancha_Shiva_Mandir_in_Pathra_(West_Bengal).jpg
[29] https://en.wikipedia.org/wiki/Bengal_roofs#/media/File:Antpur_Radhagovindjiu_Temple.jpg
[30] tps://upload.wikimedia.org/wikipedia/commons/6/6a/Radha_Govinda_Temple_Bishnupur%2C_Bankura%2C_Westbengal%2C_India.jpg
[31] https://en.wikipedia.org/wiki/Architecture_of_Bengal#/media/File:Ramchandra_Mandir,_Guptipara,_Hooghly,_West_Bengal,_India.jpg
[32] https://commons.wikimedia.org/wiki/File:Ekratna_Temple_of_Gopinath,_Khard_Radhakantapur.jpg
[33] https://en.wikipedia.org/wiki/Bengal_roofs#/media/File:Ananta_Basudeba_Temple1.JPG

flat roof, known as *chandni*. The Rupesvara temple in Kalna and the Raghunatha temple in Bardhaman are good examples of the *Dalan* roofing style. Sources[34]: [35]

7. *Rasmancha*

The *Rasmancha* used to be the centre of the autumn festivals, and it is generally octagonal, with arches opening on every side and roofs having eight turrets situated around a large central tower. Source[36]

Figure 35 Rasmancha Roof - Hindu ritual platform, the Rasmancha, Bishnupur, c. 1600

8. Rajput architecture

Figure 36 Left: Antpur Chandimandap; Middle: An 18th-century Rajput painting by the artist Nihâl Chand; Right: The Naulakha Pavilion in Lahore Fort, Pakistan, features a Do-Chala roof originating in Bengal.

Bengal roofs are dome-shaped roofs with drawn-down corners associated with the late Mughal and Rajput architecture of Northern India. It is believed that stone roofs of this type did not emerge until the 16th century and can be traced back to rural models with straw or reed roofs in the rainy regions of Bengal. Sources[37]:[38]:[39]

9. Islamic - Mughal and Rajput architecture

Figure 37 Left: A multi-domed Sultanate-era mosque; Middle: Mughal-era domes in Murshidabad; Right: Choto Sona Mosque (around 1500)

Indo-Islamic architecture in Bengali architecture can be seen from the 13th century, but before the Mughals have usually strongly reflected local traditions. The oldest surviving mosque was built during

---

[34] https://commons.wikimedia.org/wiki/File:Dalan_Temple_inside_Kaviraj_Bari_at_Mankar_situated_in_Purba_Bardhaman_district.jpg
[35] https://en.wikipedia.org/wiki/Architecture_of_Bengal#/media/File:Asiatic_Society_Bangladesh.jpg
[36] https://en.wikipedia.org/wiki/Architecture_of_Bengal#/media/File:Rasmancha,_Bishnupur.JPG
[37] https://en.wikipedia.org/wiki/Bengal_roofs#/media/File:Antpur_ChandiMandap.jpg
[38] https://en.wikipedia.org/wiki/Rajput#/media/File:Nih%C3%A2l_Chand_001_cropped.jpg
[39] https://en.wikipedia.org/wiki/Vernacular_architecture#/media/File:Naulakha_Pavilion_in_Lahore_Fort.jpg

the Delhi Sultanate. The mosque architecture of the independent Bengal Sultanate period (14th to 16th centuries) represents the most important element of the Islamic architecture of Bengal. This distinctive regional style drew its inspiration from the indigenous vernacular architecture of Bengal, including curved chala roofs, corner towers and complex floral carvings. Sultanate-era mosques featured multiple domes or a single dome, richly designed mihrabs and mimbars[40] and an absence of minarets. While clay bricks and terracotta were the most widely used materials, the stone was used from mines in the Rarh region. The Sultanate style also includes gateways and bridges. The style is widely scattered across the region. Source[41],[42],[43]

10. Mughal Bengali

Mughal Bengal saw the spread of Mughal architecture in the region, including forts, havelis, gardens, caravanserais, hammams and fountains. Mughal Bengali mosques also developed a distinct style. Dhaka and Murshidabad were the hubs of Mughal architecture. The Mughals copied the do-Chala roof tradition in North India. Source[44]:

Figure 38 Imperial Delhi Moti Masjid

11. Bungalow -The vernacular bungalow originates in Bengal, Bangladesh. Source[45]: A wooden bungalow that has served as Momin Mosque since 1920. Momin Uddin Akon decided to build the mosque with wood, as most of the houses (if

Figure 39 Left: Bungalow in Bengal, Bangladesh; Right: Momin Mosque since 1920

not all) in the village (located in the district of Pirojpur, SW Bangladesh) are made of woo. The Department of Archaeology recognized it as a national heritage and listed it under Momin Mosque for its care and protection. Source[46]:

---

[40] a short flight of steps used as a platform (a pulpit) by a preacher (usually an Imam) in a mosque.
[41] https://en.wikipedia.org/wiki/Architecture_of_Bengal#/media/File:Faridpur_PatrailMoshjid_MG_2977.jpg
[42] https://en.wikipedia.org/wiki/Architecture_of_Bengal#/media/File:Faridpur_PatrailMoshjid_MG_2977.jpg
[43] https://en.wikipedia.org/wiki/Architecture_of_Bengal#/media/File:Chapai_ChotoSonaMashjid_MG_5054.jpg
[44] https://en.wikipedia.org/wiki/Bengal_roofs#/media/File:Reminiscences_of_Imperial_Delhi_Moti_Masjid_within_the_Palace.png
[45] https://en.wikipedia.org/wiki/Vernacular_architecture#/media/File:Guest_house_in_Sylhet_(01).jpg
[46] https://en.wikipedia.org/wiki/Architecture_of_Bengal#/media/File:Momin_Mosque_after_restoration.jpg

Chapter 4

Himalayan States Stingless Bees[47] and Vernacular Structures
State of Assam

Figure 40 The Himalayan states cover Nepal, Bhutan, Assam, Nagaland and North Bangladesh.

Tiwa is an ethnic group mainly inhabiting the states of Assam and Meghalaya in northeastern India.

Figure 41 Left: Plains Tiwa's Hut; Right: A Hill Tiwa house

They are also found in some Arunachal Pradesh, Manipur and Nagaland areas. They are recognized as a Scheduled tribe within the State of Assam. Source[48]

The Hills Tiwas live in the westernmost areas of Karbi Anglong district (Assam) and the Northeastern corner of Ri-Bhoi district (Meghalaya). Plains Tiwas live on the flat lands of the Southern bank of the

---

[47] Stingless bees of the genus Lepidotrigona are very common but confined to the north-eastern states of India (Viraktamath and Rojeet 2021). They are known with various names in different states of North East India. They are called as "Khoi-te" in Mizoram; "Dkai/ Ngap Syor/ Ngap ryngkai/ Ngap hamang/ Ngap khyndew/ U-Ngap Rpiang /Mengkari/ Bengbengkil" in Meghalaya; "Putka" in Sikkim; "Khoining-Khoi/ Raona/ Laona/Kalei-Leiva/ Ora-Lei/ Shalu/ Laichih/ Balep/ Mikza" in Manipur; "Jahrah-Konying/ Rhontso-Tsak/ Helia-Dui and Nyangkit" in Nagaland; "Sung Bihin Kun Kuni Mou" In Assam; "Piya Ka/ Piya Kusumui" in Tripura and in Arunachal Pradesh, they are known as Ter/ Tar in local dialect.

[48] https://en.wikipedia.org/wiki/Tiwa_people_(India)

Brahmaputra valley, mostly in Morigaon, Nagaon, Kamrup (Rural) and (Metro), Sibsagar, Lakhimpur and Dhemaji districts. Source[49]

Meghalaya, NE India

The Khasi people are an ethnic group of Meghalaya in north-eastern India with a significant population in the bordering state of Assam and certain parts of Bangladesh. Source[50]

Figure 42 The royal seat of Khyrim at Smit

Native bee names related to the people and locality:

khasiana, *Melipona* **Pugh** 1947 [Nom nud.][51] (Rasmussen, 2008) = **Khasi** (plural **Khasis** or **Khasi**) - A member of an Indian tribe, the majority of whom live in **Meghalaya**, India + ana (see suffix –**ana**)

Tetragonula kyrdemkulaiensis Viraktamath and Rojeet sp. n. Etym: This species is named after the place where it is collected. Holotype: Male: **Meghalaya**: Kyrdemkulai (25.44° N, 91.47° E, Altitude 679 m a.s.l.), 20. ii. 2019, Coll. Rojeet T. deposited at UASB.
Remarks: This stingless bee was found in a subterranean nest at Kyrdemkulai, Ri-Bhoi District of Meghalaya, in January 2019. The nest was located at a depth of 90 cm in the soil, with an entrance burrow (116 cm in length) running horizontally.

Tetragonula srikantanathi Viraktamath sp. n. (3) This species is named in honour of Mr. Srikanta Nath. Holotype: Female: **Tripura**: Salema (24.01° N 91.83° E, Altitude 89 m a.s.l.), 8. xi. 2017, Coll. S. Viraktamath, deposited at UASB.
Remarks: Mr. Srikanta Nath, Assistant Director of Agriculture of the Tripura State Agriculture Department, searched the nest and helped study its structure.

Tetragonula bengalensis (Cameron 1897)
Trigona bengalensis Cameron 1897: 143–144: Lectotype (OUMNH, worker): examined, "Trigona / bengalensis / Cam.". In addition, "LECTOTYPE Trigona bengalensis Cameron Design. C Rasmussen 2013"; **Type locality**: INDIA, **West Bengal**, 26 km N of Kolkata (formerly Calcutta) on the east bank of the Hooghly river (also known as Hugli) (sometime during 1872–1886 or 1893) [ca. 22.68°N, 88.38°E].

Figure 43 Tetragonula bengalensis (Cameron 1897)

---

[49] https://tiwatribe.blogspot.com/
[50] https://en.wikipedia.org/wiki/Khasi_people
[51] The name is unavailable for taxonomic purposes, as no description or definition of the taxon was provided (ICZN, art. 13.1.1) (common name (nagp hamang)

Sikkim, Himalayas, India

Smack in between Nepal and Bhutan, Sikkim (/ˈsɪkɪm/; Nepali pronunciation: [ˈsikːim]) is a state in Northeastern India. It borders the Tibet Autonomous Region of China in the north and northeast, Bhutan in the east, Province No. 1 of Nepal in the west and West Bengal in the south. Sikkim's capital and largest city is Gangtok, notable for its biodiversity, including alpine and subtropical climate. Our interest in this cold region sparked interest, knowing that there are stingless bees in Nepal and Bhutan. True enough, there is a record of a native bee in that locality:

Figure 44 Lepidotrigona arcifera Photo Claus Rasmussen

Lepidotrigona arcifera (Cockerell, 1929)
Trigona arcifera Cockerell 1929c: 591-592: Holotype (BMNH 17b.1081). The ventralis species group (taxonomy);
Type locality: INDIA "**Testa bridge, Himalayas**, India, 10.1.97 (Sladen)" (worker);

The Upper Dzongu forest block and its villages are located on the other side of the Rungyung Chu River in Sikkim. To access the Upper areas of Dzongu, one must cross the massive Rungyung Chu riverbed. The vast riverbed can be crossed through a beautiful suspension Bridge in the Mantam area. Some upper Dzongu area villages include Tingvong, Kusong, Sakyong-Pentong and Lingthem.

Figure 45 Village dwellings in The Upper Dzongu forest block in Sikkim

Motorable roads end at Lingzya in Upper Dzongu after Tingvong. One has to trek to other villages lying beyond. Source[52]

We have covered some of the Island vernacular structures in the Indo-Malayan Ecozone, and now we look at some highland structures in The Himalayan states. The author's previous works[53] in those regions were briefly covered, but the approach here is more on indigenous structures to be applied for

---

[52] https://en.wikipedia.org/wiki/Upper_Dzongu_Forest_Block
[53] We have covered Meliponiculture in Nagaland (contributions from Bode Shuya and Dr. Thungben Yanthan) and Bhutan (contributions from Drupka Nymgyal) in the Handbook of Meliponiculture in 2016 and Nepal (Contribution from Dr. Chet Bhatta) in the Atlas of Meliponiculture in 2018. In the book on "Geometry of Meliponine Brood Cells, we covered *Tetragonula kyrdemkulaiensis* in Kyrdemkulai (Contributed by Dr. Shashidhar Varaktamath).

bee housing. The round-ended tribal long structures in Nagaland (NE India) have similarities to The Shan tribal Home of Burma (See Figure 83 Shan tribal House) and also the Kadazan long house of Sabah (See Rumah Panjang Kadazan Sabah or Sabah Kadazan Longhouse in Volume 1 Part 1). We will scrutinize the Himalayan States' traditional architecture and see if it can be applied to indigenous ethnic bee housing.

The recent finding of *Lepidotrigona* sp. nov. 2022 in Sikkim States is as follows:

Lepidotrigona sikkimensis Viraktamath sp. n. **Etym.**: The species name is derived based on the state's name, "Sikkim", where this species was collected. **Holotype**: Male: Sikkim: Mamley (27.1907° N, 88.3726° E), 09. xii. 2017. leg. S. Viraktamath, deposited at UASB. **Paratypes**: Eight males and 70 workers with the same collection data deposited at UASB; one worker paratype will be deposited at ZSIK.

Nagaland

The structures with Buddhist elements, like stupas and pagodas, have similar Tibetan styles. The ethnic tribes in Nagaland, however, have their unique construction. Unlike the tropical Islands, these homes are built to keep warm with thick thatch roofs. Interestingly, one type of construction caught my attention because of its similarity to Peru's. The Sümi or Sema Naga traditional dwelling is compared to this by the Aguarano people of Peru (See Figure 46).

Figure 47 The ethnic tribes in Nagaland, however, have a unique construction.

Figure 46 Similarities in the 11th-century Nordic battle preparation camp and the construction of the Aguarano home in Peru.

Figure 48 Beehouse in the design of the Sema Naga house.

The other uncanny similarity was a Nordic battle preparation camp in the 11th century (the precursor of the Battle of Hastings 1066 - something I recall from my school's history lessons 50 years. ago). The battle preparation was a congregation of Viking tribes for the invasion of England in retaliation for the St. Bryce Day massacre.

We see how the Buddhist influence flowed through the Himalayan states. Passing some head hunter tribes in Nagaland and Northern Assam, the influence had to skip and hop over to Burma. The pointed spires of Stupas in Nepal and Bhutan disappeared in Nagaland, where the two bamboos forming the gable were prolonged beyond the roof to form horns called the *tenhaku - ki* (snail horns). The following images examine Himalayan head-hunters, the Sema Naga Tribal House (Figure 48). These structures

Figure 49 Tribal structures in Nagaland structures are very similar to Viking houses.

are very similar to Viking houses. Although Vikings may have a more diverse history, they carried Southern European and Asian heritage. It inspired me to dish out this design.

Meliponiculture in Nagaland

Figure 50 Meliponiculture in Nagaland by Bode Shuya

Mizoram State in NE India

The Mizo people (Lushai: Mizo hnam) are an ethnic group native to the Indian state of Mizoram and neighbouring regions of Northeast India. The term covers several related ethnic groups or clans inside the Mizo group. The term Mizo is derived from two Mizo words: Mi and Zo. Mi in Mizo means "person" or "people". The term Zo has two meanings. According to one view, Zo means 'highland' or perhaps 'remote'.

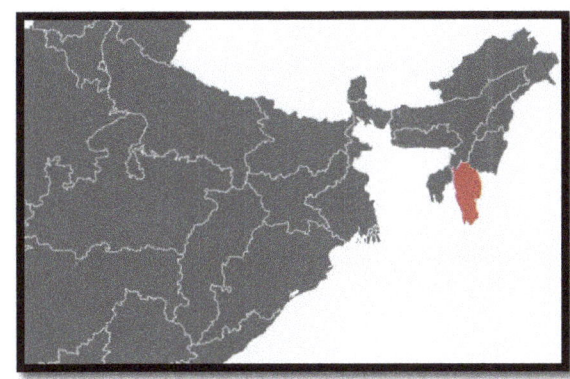

Map 1 Location of Thenzawl (Mizoram, 23.2808° N, 92.7741° E) NE India.

Another meaning is "cool" or "crisp" (i.e., a sense/feeling of cool and refreshing air/environment of higher altitude. The present Indian state of Mizoram (which means "Mizoland") was historically called the Lushai Hills or Lushai District. The Lushai Hills area was defined as an excluded area during the British Raj and as a district of Assam in independent India.

Figure 51 Top left: Local Thenzawl architecture and Traditional structures in Mizoram, NE India.

Bee species were discovered in Thenzawl, Mizoram state (See Map 1).

1. ***Lepidotrigona thenzawlensis*** Viraktamath and Rojeet sp. n. **Etym**.: This species is named after the place Thenzawl from where the samples were collected. **Holotype**: Male: Mizoram: Thenzawl (23.2808° N, 92.7741° E, Altitude 783 m a.s.l.), 02. ix.2020, leg. Rojeet T. deposited at UASB. **Paratypes**: 12 males and 25 females with the same collection data deposited at UASB; one female paratype will be deposited at ZSIK.

2. ***Lepidotrigona rajithae*** Viraktamath and Rojeet sp. n. **Etym.***:* The species name is derived from the Sanskrit word "Rajitha", meaning "impressive", referring to large male bees. **Holotype**: Male: izoram: Thenzawl (23.2808° N, 92.7741° E, Altitude 783 m a.s.l.), 02.ix.2020, leg. Rojeet T. deposited at UASB. **Paratypes**: 12 males and 25 females with the same collection data deposited at UASB; one female paratype will be deposited at ZSIK

3. ***Lepidotrigona amruthae*** Viraktamath and Rojeet sp. n. **Etym**.: The species name is derived from the Sanskrit word "Amrutha", meaning "nectar of immortality". The name refers to the high medicinal value of the honey produced by stingless bees. **Holotype**: Male: Mizoram: Thenzawl (23.2808° N, 92.7741° E, Altitude 783 m a.s.l.), 12.x. 2020, leg. Rojeet T. deposited at UASB. **Paratypes**: 10 males and 28 females with the same collection data deposited at UASB; one female paratype will be deposited at ZSIK.

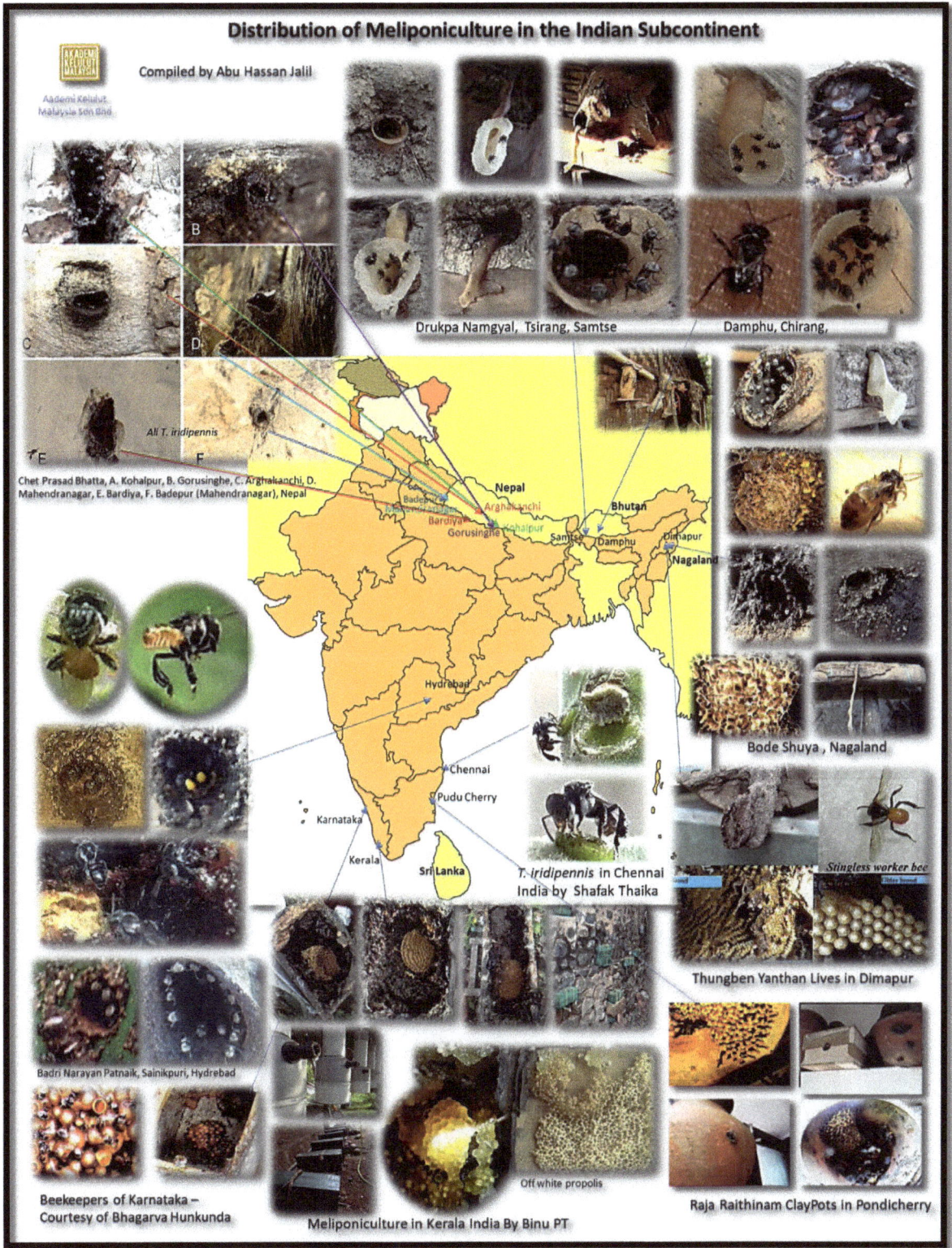

Figure 52 Distribution of Meliponiculture in the Indian Subcontinent

Meliponiculture is also practised in Tamil Nadu, Maharashtra, Gujarat, Assam, Meghalaya, Manipur, Assam, Sikkim and Andaman & Nicobar Islands besides the states that are shown in this chart (Pers.. Comm Viraktamath, Feb. 2023)

Goal Ghar of Nepal

Figure 53 Inspired by the Typical Nepali "Goal Ghar" or roundhouse

Stingless Bees in Bhutan[54]

Figure 54 Drukpa Namgyal Lives in Damphu, Chirang, Bhutan

---

[54] More data in Nidup, T. (2021, April 26) Retrieved from Journal of Threatened Taxa

Chapter 5

Sri Lanka

Figure 55 Vernacular architecture in Sri Lanka and South India.

There was limited bee data on Sri Lanka, but the vernacular architecture is interesting and easily available. Sri Lanka is an island towards the west end of the Indo-Malayan Eco-zone.

Tetragonula iridipennis (Smith 1854)

Trigona iridipennis Smith 1854: 413–414: Lectotype (BMNH 17b.1114, worker): examined, "Type" (orange border), "iridipennis / Type Sm.", "B.M. TYPE / HYM. / 17B.1114", "TRIGONA / iridipennis / TYPE. Smith.", "Ceylon" (reverse side "53 / 23"). In addition, "LECTOTYPE Trigona iridipennis Smith Design. JS Moure 1961 / C Rasmussen 2013"; **Type locality**: SRI LANKA, Central Province, Kandy [ca. 7.27°N, 80.60°E, ca. 467m a.s.l.].

Tetragonula praeterita (Walker 1860)

Trigona præterita Walker 1860: 305–306: Lectotype (BMNH 17b.1185, worker): examined, "63 / 52", "Type" (green border), "præterita / W.", "Trigona", "B.M. TYPE / HYM. / 17b.1185" (taxonomy). In addition, "LECTOTYPE Trigona praeterita Walker Design. JS Moure 1961 / C Rasmussen 2013"; **Type locality: SRI LANKA**[55].

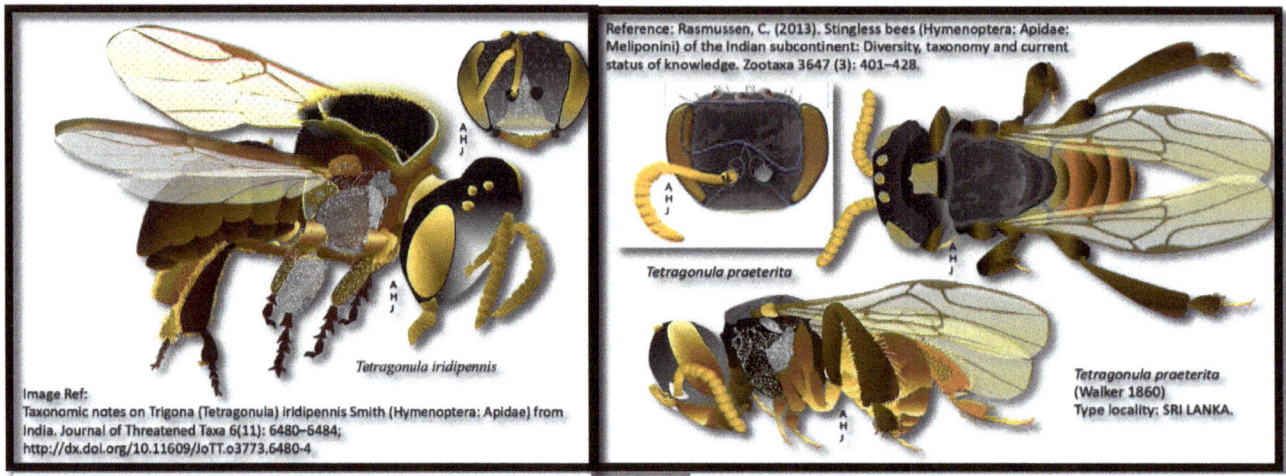

Figure 56 Tetragonula iridipennis (Smith 1854) Figure 57 Tetragonula praeterita (Walker 1860)

Sinhalese people[56] (Sinhala: සිංහල ජනතාව, romanized: Sinhala Janathāva) are an Indo-Aryan ethnolinguistic group native to the island of Sri Lanka.

Etymology

Sinhala is derived from Sanskrit siṃhá, literally "lion" with the suffix -la, together meaning "abode of lions", referring to the prevalence of lions in earlier Sri Lankan history.

The form of Buddhism in Sri Lanka is known as Theravada (school of elders)[57].

Figure 58 Patthirippua at Sri Dalada Maligawa Kandy, Sri Lanka (The Temple of the Tooth)

---

[55] *Lisotrigona cacciae* is also reported from Sri Lanka (Please see Karunaratne et al, 2017) First record of a tear-drinking stingless bee *Lisotrigona cacciae (Nurse)(Hymenoptera: Apidae: Meliponini), from the central hills of Sri Lanka. Article in Journal of the National Science Foundation of Sri Lanka · March 2017*

[56] https://en.wikipedia.org/wiki/Sinhalese_people

[57] https://en.wikipedia.org/wiki/Temple_of_the_Tooth

Indigenous people of Sri Lanka

The Vedda[58] (Sinhala: වැද්දා [ˈvædːaː], Tamil: வேடர் (Vēḍar)), or Wanniyalaeto, are a minority indigenous group of people in Sri Lanka who, among other sub-communities such as Coast Veddas, Anuradhapura Veddas and Bintenne Veddas, are accorded indigenous status. The Vedda minority in Sri Lanka may become completely assimilated.

The original religion of Veddas is animism. The Sinhalized interior Veddahs follow a mix of animism and nominal Buddhism, whereas the Tamilized east coast Veddahs follow a mix of animism and nominal Hinduism with folk influences among anthropologists.

Figure 59 Left: Wanniyala-Aetto Village Right: Veddahs (wild men), Ceylon.

Today, the Vedda live scattered in tiny settlements in the Hunnasgiriya hills in central Sri Lanka up to the coastal lowlands in the island's east. However, long before Indo-Aryans –now the dominant Sinhalese-Buddhist people – came to Sri Lanka from India around 543 BCE, the Vedda lived all around the island.

Sri Lankan Malays

Sri Lankan Malays[59] (romanized: Ilaṅkai Malāi Makkal are Sri Lankans with full or partial ancestry from the Indonesian Archipelago, Malaysia, or Singapore. In addition, people from Brunei and the

Sri Lankan Mud House

Figure 60 Mud House in NW Sri Lanka

Similarities between Sri Lanka Architecture and South India and their relation with bees in Kerala and Karnataka up to Pondicherry were compared.

The Meghalaya Cleanest House 2022

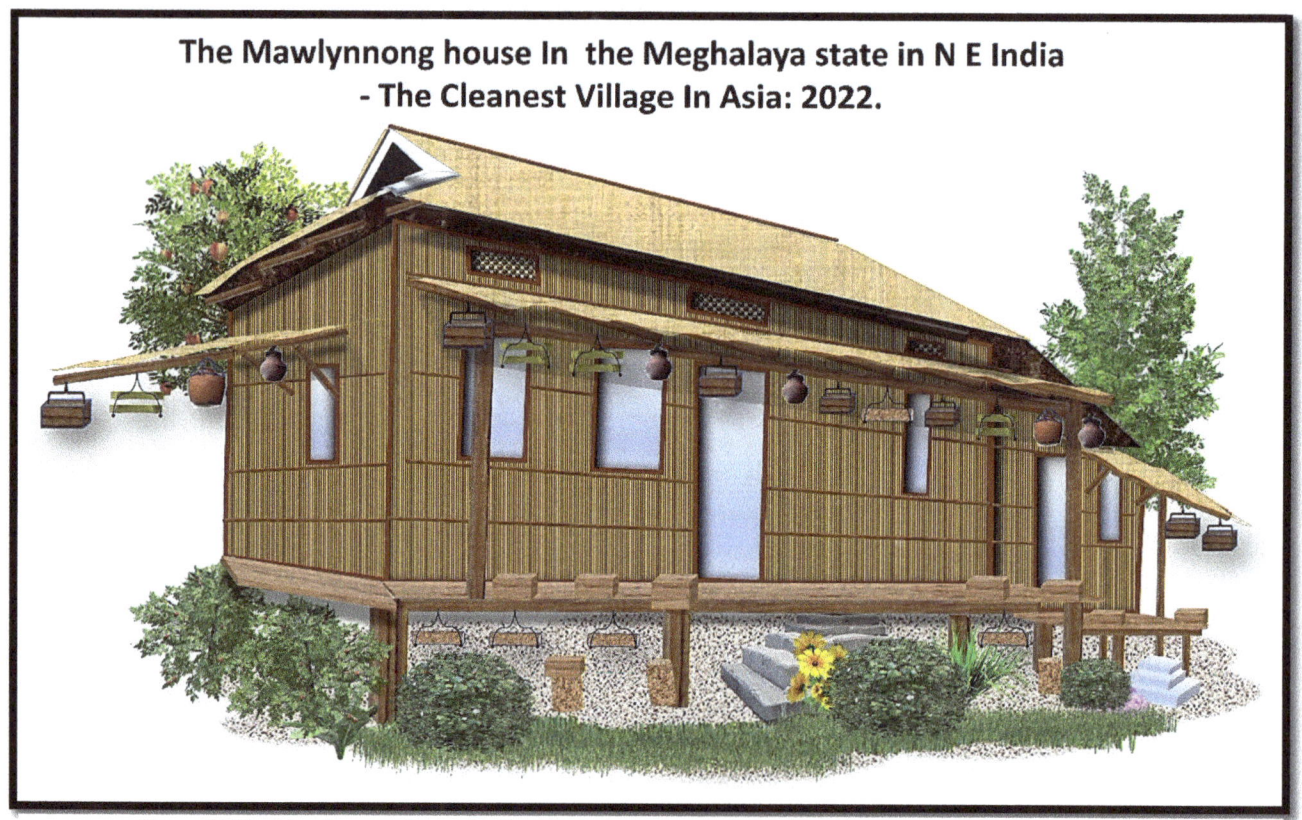

Figure 61 The Mawlynnong house In the Meghalaya state in N E India - The Cleanest Village In Asia: 2022.

VOLUME 2, PART 2
~ INDO-CHINA ~

Preambles

People of Southeast Asia

Malay, Orang Melayu ("Malay People"), are any member of an ethnic group of the Malay Peninsula and portions of adjacent Southeast Asia, including the east coast of Sumatra, the coast of Borneo, and smaller islands between these areas. The Malays speak various dialects belonging to the Austronesian (Malayo-Polynesian) family of languages.

The Malays were once probably a people of coastal Borneo who expanded into Sumatra and the Malay Peninsula due to their trading and seafaring way of life. That this expansion occurred only in the last 1,500 years is indicated by the fact that the Malay group's languages are all very much alike, though very divergent from the languages of other peoples of Sumatra, Borneo, and other neighbouring lands. In the early 21st century, the Malays constituted about half of the population of Peninsular Malaysia (West Malaysia) and one-eighth of East Malaysia (Sarawak and Sabah). [Fay-Cooper Cole - American anthropologist]

Thailand - Chinese of Thailand

Thailand has attracted large numbers of immigrants from neighbouring countries since the mid-19th century, owing to the expansion of the Thai economy and political upheavals elsewhere in Asia. The largest number of immigrants by far have come from China, constituting a significant minority in Thailand.

In the early 1900s, about one-seventh of the country's population was identifiable as Chinese, but by the early 21st century, roughly one-tenth of the population still recognized its Chinese ancestry. The overwhelming majority of people of Chinese descent (Thai: *luk cin*) in contemporary Thailand have been assimilated into Thai culture, largely by adopting Standard Thai as their primary, or even exclusive, language and becoming Theravada Buddhists. These assimilated Chinese are known in English as Sino-Thai.

Upland-dwelling peoples (also known as "hill tribes") such as the Karen, Hmong, Yao, Lahu, Lisu, and Akha also follow distinctive traditions that set them apart from the country's Tai-speaking majority. In the past, such people were considered by the Thai to be people of the forest, and this association has continued to shape the popular image of upland communities in the 21st century. Most upland people at one time followed local religious

Figure 62 Thailand upland-dwelling peoples (also known as "hill tribes")

traditions. While some have become Buddhists, more have converted to Christianity, distinguishing them from most of the population.

The Chinese brought miniature landscape art of 'Pen Jing' on a tray. (See the last chapter herein.) Penjing is an ancient Chinese art form of creating landscape scenes on a miniature scale.

Malays, upland peoples, and new immigrants.

Not all people living within the borders of Thailand have been fully integrated into the national community. Malay-speaking peoples comprise most of the country's four southernmost provinces. Because this region constituted a separate Malay sultanate until the late 19th century, and because its inhabitants have a distinct linguistic identity and religious heritage (as practitioners of Islam), some of the residents of the area have supported movements seeking greater autonomy or even independence from the predominantly Buddhist and Tai-speaking rest of the country.

Malay culture has been strongly influenced by the cultures of other areas, including Thailand, Java, and Sumatra. The influence of Hindu India was historically very great. The Malays were largely Hinduized before they were converted to Islam in the 15th century.

Many Malays are rural people living in villages rather than towns. Traditional houses are built on pilings that raise them four to eight feet off the ground, with gabled roofs made of thatch; more affluent houses have plank floors and tile roofs. Much of the Malay Peninsula is covered by jungle, and the villages, with populations from 50 to 1,000, are located along rivers and coasts or on roads.

Traditionally, Malay social organization was somewhat feudal, with a sharp division between nobility and commoners. The head of a village was a commoner, but the chief of the district, to whom the village head reported, was a member of the nobility. Since the late 20th century, the nobility has been replaced by appointed and elected officials subject to a parliament and other elected bodies, although class distinctions have persisted. With rapidly accelerating rural-to-urban migration, many Malays have left their villages to settle in cities, towns, and suburbs, where they now work in virtually every industry.

Marriage and inheritance are governed by Sharīʿah (Islamic law)[60]. Parents have traditionally arranged marriages. The typical household consists of the husband and wife and their children.

Concerning Malay origin, the ethnic group is called the Urak Lawai. Urak Lawoi (Malay: Orang Laut; Thai: อูรักลาโว้ย; RTGS: Urak Lawoi) are an Aboriginal Malay people residing on the islands of Phuket, Phi Phi, Jum, Lanta, Bulon and on Lipe and Adang, in the Adang Archipelago, off the western

---

[60] Source: https://www.britannica.com/topic/Malay-people

coast of Thailand. They are known by various names, including Orak Lawoi', Lawta, Chao Tha Le (ชาวทะเล), Chao Nam (ชาวน้ำ), and Lawoi[61].

A related ethnic group is the Moken (also Mawken or Morgan; Burmese: ဆလုံ လူမျိုး; Thai: ชาวเล, romanized: Chao le, lit. 'sea people' – see also Sama Bajao Factor in Chapter 10)) are an Austronesian people of the Mergui Archipelago, a group of approximately 800 islands claimed by both Myanmar and Thailand. Most of the 2,000 to 3,000 Moken live a semi-nomadic hunter-gatherer lifestyle heavily based on the sea, though this is increasingly under threat.

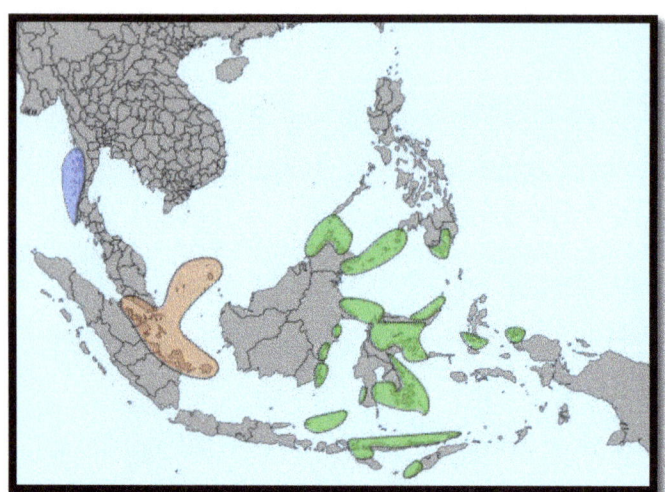

Figure 63 Distribution of three different peoples usually called "Sea Nomads" or "Sea Gypsies": Blue: Moken Orange: Orang Laut Green: Sama-Bajau

Aside from ancestor worship, the Moken have no religion[62]. Their egalitarian society follows their ancestral worship as they regularly present supernatural beings with food offerings. Their knowledge of the sea enables them to live off its fauna by using simple tools such as nets and spears to forage for food, which allows them to impact the environment more minimally than other more intensive forms of subsistence.

---

[61] Source: https://en.wikipedia.org/wiki/Urak_Lawoi
[62] https://en.wikipedia.org/wiki/Moken

Introduction to Part 2

Malay architecture influence is widespread in the ASEAN countries and even in ethnic island traditional houses. Besides traditional bee hive box roofing, this Part looks at the overall bee housing, racks, shacks and sheds globally. It will touch mainly on the Vernacular Architecture of Indo-China as well as compare rake and fascia boards in Indo-China.

'Out of the box but within the sphere' is given to describe something out of the norm but still within the bounds of reality. This Part examines the Fusion of Traditional architecture with Sino-Portuguese and European Fusion but brings to mind colonial influence on local Asian flare. Then there are the Himalayan States Vernacular Structures and the Diversity of Indo-Malayan Vernacular Architecture.

Continuing from the ancestor veneration episodes, I look at cultures that keep the bones and skulls of their ancestors. Some are for worship, and some serve to keep the spirits alive. It brings us to the Himalayan states because it includes the Nagaland people who were head hunters at one time, with their peculiar traditional houses with intriguing façades and tribal ornaments on their roofs.

The Totem Poles Avenue brings us to architecture that keeps the 'Head Hunter' vernacular accent, mostly in the Dayaks of Borneo and indigenous peoples like the Asmat of Papua. Henceforth, this Part gathers the most recent headhunting news as late as 2010 in Central Kalimantan.

Figure 64 Unique Architecture of Rong House

Rong house is a typical structure on stilts. This is the community house, like the community house of the Kinh people, used as a place for meeting, exchanging and addressing the villagers in the Central Highlands. The house also welcomes guests (according to Ba Na custom) besides guests of families or villages. Rong houses are only in ethnic villages like Gia Rai and Ba Na… in North Highlands, especially in Gia Lai and Kon Tum. (As reported by Ho Lam)

Source: https://www.dalattrip.com/fly/rong-house-soul-of-ethnic-villages-in-central-highlands/

Chapter 6

Indo-China Vernacular Architecture

In the Philippines and Thailand, it was observed that not many beekeeping practices were connected with vernacular architecture other than the Tausug Architecture of ancestral homes in Panay Island, Philippines, unlike what was expected in other countries in the region where there may be some indigenous architecture with thatch roofing applied in Meliponiculture housing.

There is a sore lack of information on Thai suburban beekeeping besides visits to Bangkok and Chiang Mai. Those in The Raja Mangala University in Bangkok and the Chiang Mai University in Chiang Mai are standard bee housing and are kept for studies. The paper Stingless Bee Beekeeping in Thailand[63] describes the standard box hives (i.e., a shoebox cuboid) in a commercial Meliponary at the southeast border with Cambodia. The hive racks are in standard shade structures as well. That aside, on the topic of vernacular architecture, this Chapter takes a closer look at Thai traditions for the benefit of those beekeepers who are into model-making for bee housing and is also a new avenue for insect tourism.

Figure 65 Traditional Thai village homes with many similarities with the Malay village house design

The village houses look similar to the Malay Kampung houses way down south on the Peninsula. The difference is the ornate fascia boards, Gable/rake carvings and the roof ridge ornaments like the end finials (spire). Interestingly, even though Buddhist Stupas and pagodas abound, the formal roof ornaments in the Chofah[64] architecture depict some form of mythical beings like the Garuda and Naga. With Buddhism originating from India, we see similar mythical depictions in the Majapahit era architecture in Java and

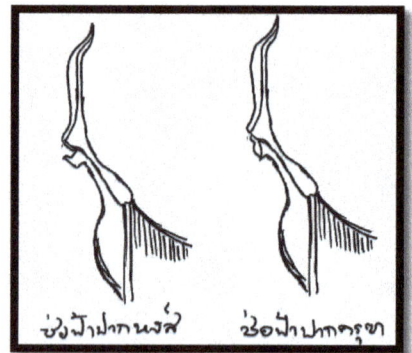

Figure 66 Two main types of Chofah: Pak Hong; Swan's tip (left) and Pak Khrut; Garuda's tip (right)

---

[63] Bajaree Chuttong et al. = Stingless Bee Beekeeping in Thailand - April 2015Bee World 91(2):41-45 DOI:10.1080/0005772X.2014.11417595

[64] Chofa (Thai: ช่อฟ้า, pronounced [tɕʰɔ̂ː.fáː]; lit. sky tassel) is a Lao and Thai architectural decorative ornament that adorns the top at the end of wat and palace roofs in most Southeast Asian countries, such as Thailand, Cambodia, Laos, and Myanmar. Source: https://en.wikipedia.org/wiki/Chofa

the Naga (dragon/serpent) in China. The Buddhist influence is heavy in Thailand's architecture and surrounding neighbours Myanmar, Laos and Cambodia.

Other creature depictions like the elephant, fish, and birds with long necks constitute the finials and ornaments. Below are some examples of oriental mythical figures that may adorn Indo-Chinese roofs.

Architecture changes as we get into Vietnam. The upsweep of the corner ridges shows some Chinese influence and French colonial influence, and the

Figure 67 Left: Oriental mythical figures that may adorn Indo-Chinese roofs; Right: Oriental influence on a Vietnamese house roof applied on a box hive with the rooster roof ornaments and upswept roof eave corners.

roof ornaments change to the rooster (French Le coq, Filipino roof ornament Sarimanok (Maranao) or Harimanok (Tagalog). In Java, Harimanok means chicken day.)

This model replicates a Vietnamese house that may be scaled down to a bee box hive roof. We have not seen Vietnamese beekeepers adorning their box hives with their traditional architecture. At the same time, there are some great box makers in Vietnam, but durability and sustainability have not been fully tested as Meliponiculture is relatively new in Vietnam. However, tourism in Vietnam is picking up. From visits to Ho Chi Minh City 7 or 8 years ago, I observed that beekeepers kept stingless bees in Bamboo hives. There are some interesting discoveries there, then[65]. The bees visit lotus flowers to die in the petals. Never had a chance to find out why, but that's another topic.

Figure 68 Filipino roof ornament Sarimanok (Maranao)

Not including temples, stupas and pagodas, the main differences among the Indo-China traditional dwellings, pavilions and palaces or administrative structures among the Indo-China countries are the

---

[65] Editors' note: [of course they do, Apis florea is all over them, gathering pollen] ... Please see my book for the COP 8 conference in Germany, where I used a photo of this...

roof ornaments, finials, carved fascia and rake and roof pitch. The basic gable and hip roofs are similar throughout.

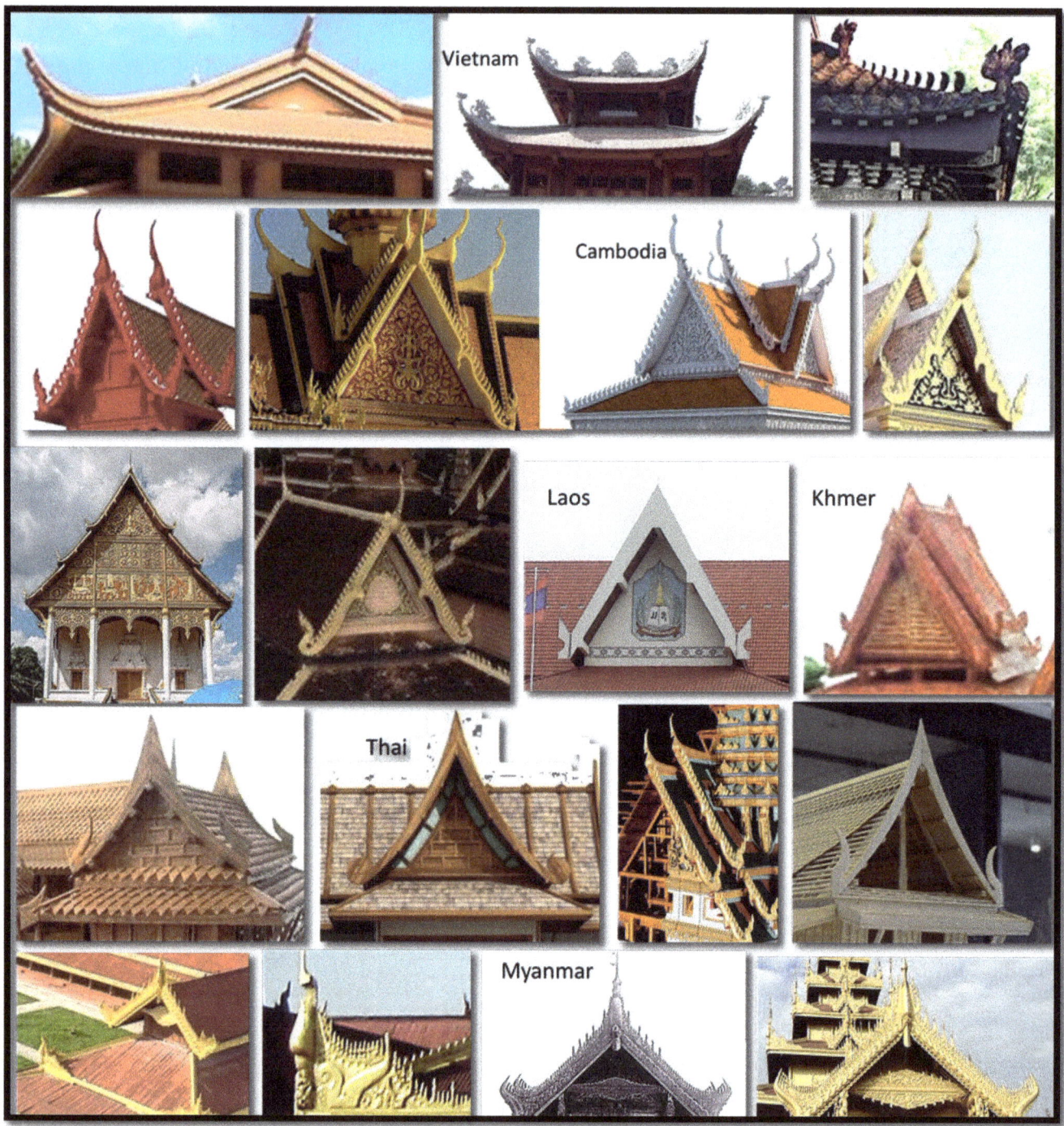

Figure 69 Comparing Rake and Fascia boards in Indo-China

Figure 70 https://upload.wikimedia.org/wikipedia/commons/5/57/Mandalay_Palace.JPG

However, the Burmese are influenced by Chinese architecture and Buddhism simultaneously. Incorporating an upsweep in the roof eaves is difficult, so the Burmese add Buddhist ornaments at the eave corners and rake boards.

Comparing Rakes and Fascia boards in Indo-China

Figure 71 Ancient Khmer Architecture influenced pediments, rake board facia and finials.

Not everything is ornamental or intricately designed. An example is the **Dai** people (pinyin: Dǎizú) refers to several Tai-speaking ethnic groups living in the Xishuangbanna Dai Autonomous Prefecture and the Dehong Dai and Jingpo Autonomous Prefecture of China's Yunnan Province. The Dai people form one of the 56 ethnic groups officially recognized by the People's Republic of China. By extension,

Figure 72 Model of a Dai bamboo house.

the term can apply to groups in Laos, Vietnam, Thailand and Myanmar when Dai means specifically Tai Yai, Lue, Chinese Shan, Tai Dam, Tai Khao or even Tai in general[66].

Another example is the Palaung (Burmese: ပလောင် လူမျိုး [pəlàʊɰ̃ lùmjó]; Thai: ปะหล่อง, also written as Benglong Palong) or Ta'ang are a Mon–Khmer ethnic minority found in the Shan State of Burma, Yunnan Province of China and Northern Thailand. In China, they are referred to as the De'ang (Chinese: 德昂族; pinyin: Déángzú also spelt Deang) people[67].

Figure 73 Left: The Palaung in the early 1900s; Right: Palaung village near Kyaukme, Myanmar (2017)

Myanmar Indigenous Architecture

Kayin Traditional Houses

Figure 75 The Sagwa Kayin House

Figure 75 The Paw Kayin House

---

[66] https://en.wikipedia.org/wiki/Dai_people
[67] https://en.wikipedia.org/wiki/Palaung_people

An Architectural Study on Kayin Traditional Homes.

Figure 76 The Bwe Kayin House

The main characteristics of Kayin traditional houses are bamboo and timber structures raised on stilt posts, with easily removable ladders and pitch roofing. The most significant feature of Kayin's house is the roof made of bamboo tiles and concave plates interlocking with convex plates. The bamboo selected for the posts is twenty or more feet long, usually four to six inches in diameter. They are set in the ground at intervals of four or five cubits (six to seven and a half feet). Holes are chopped through this large upright at six to eight feet above the ground, and pins are thrust through which bamboo girders of the same size are fastened employing withes[68]. The floor is made of large bamboo, split and flattened out and secured to the joists utilizing withes of the same material. It is six or eight feet above the ground, springy, and seamed with cracks, through which rubbish and wash water may be disposed of. The old Kayin carpenters constructed these traditional houses.

Shan Tribal House

The ideal Shan house has a round-ended roof, referred to as *hern long toob mon* (Figure 83); however, this form of the house is rarely seen today. A reference to this type of house could be found in some old photographs Western scholars took during the nineteenth century. An early development phase shown in the same figure's back row is identified as a two-roof composition, with the main house with a round-ended roof and

Figure 77 Shan Tribal House Redrawn from Oranratmanee, R. (2018)

a smaller one next to it. This type evolved due to the demand for the separation of the kitchen and the need for additional sleeping places for sub-families of the sons who, by Shan custom, reside in the same house after marriage.

---

[68] plural noun of **withy** - another term for **osier**.
a tough, flexible branch of an osier (shoot of willow) or other willow, used for tying, binding, or basketry.
"it is fixed with withies tied to the common rafters"

Inle Lake, Nyaungshwe Township of Shan State

Inle Lake is a freshwater lake in the Nyaungshwe Township of Shan State, Part of Shan Hills in Myanmar (Burma).

Figure 78 Stilt houses around Inle Lake, some with unique vernacular architecture

The people of Inle Lake (see Intha) live in four cities bordering the lake, in numerous small villages along the lake's shores, and on the lake itself. The entire lake area is in Nyaung Shwe township. The population consists predominantly of Intha, with a mix of other Shan, Taungyo, Pa-O (Taungthu), Danu, Kayah, Danaw and Bamar ethnicities[69]. Most are devout Buddhists and live in simple houses of wood and woven bamboo on stilts. They are largely self-sufficient farmers; some may even have a floating tomato garden[70]. This may be an ideal situation to utilize bees for pollination.

The Tenasserim Hills[71] or **Tenasserim Range** (Burmese: တနင်္သာရီ တောင်တန်း, [tənìɰ̃θàjì tàʊɰ̃dáɰ̃]; Thai: ทิวเขาตะนาวศรี, Malay: Banjaran Tanah Seri/Banjaran Tenang Sari) is the geographical name of a roughly 1,700 km long mountain chain, Part of the Indo-Malayan Mountain system in Southeast Asia. It is also the parent range to The Titiwangsa Range[72] (Malay: Banjaran Titiwangsa, بنجرن تيتيوڠسا, pronounced [ˈband͡ʒaˈran titiwaŋˈsa]), also known as "Banjaran Besar" (Big Range) by locals, is the chain of mountains that forms the backbone of the Malay Peninsula. The northern section of the range is in southern Thailand, where it is known as the Sankalakhiri Range (Thai: ทิวเขาสันกาลาคีรี, pronounced [tʰīw kʰǎw sǎn.kāːlāːkʰīːrīː]). Geographically, Myanmar, Thailand and Peninsula Malaysia are united by this mountain range.

---

[69] https://en.wikipedia.org/wiki/Inle_Lake
[70] https://en.wikipedia.org/wiki/Inle_Lake#/media/File:Inle_Lake,Floating_Garden.JPG
[71] https://en.wikipedia.org/wiki/Tenasserim_Hills
[72] https://en.wikipedia.org/wiki/Titiwangsa_Mountains

Records of Meliponines on the Tenasserim Range (Rasmussen 2008)

Geniotrigona thoracica (Smith, 1857)
   Trigona thoracica Smith 1857: 50: Bingham 1897: 561, 564 (distribution, key to species, taxonomy); MYANMAR, Tenasserim;

Homotrigona lutea (Bingham, 1897)
   Melipona lutea Bingham 1897: 559, 560, 564: Type (BMNH 17b.1122) (distribution, illustration, key to species, taxonomy; Type locality: MYANMAR "Northern Tenasserim; Karennee" (worker); Homotrigona lutea (Bingham, 1897)

Lepidotrigona terminata (Smith, 1878)
   Trigona terminata Smith 1878: 169: Type (BMNH 17b.1100). The terminata species group (taxonomy); Type locality: MYANMAR no precise locality, presumably "Maulmain, Tenasserim Provinces" (worker);

Lepidotrigona ventralis (Smith, 1857)
   Trigona ventralis Smith 1857: Bingham 1897: 560, 562-563 (distribution, key to species, taxonomy); MALAYSIA, Borneo, Malacca; MYANMAR, Tenasserim

Lophotrigona canifrons (Smith, 1857)
   Trigona canifrons Smith 1857: Bingham 1897: 560, 562 (distribution, key to species, taxonomy); INDONESIA, Sumatra; MALAYSIA, Borneo; MYANMAR, Tenasserim;

Tetragonilla atripes (Smith, 1857)
   Trigona atripes Smith 1857: Bingham 1897: 560, 561 (distribution, key to species, taxonomy); MALAYSIA, Malacca; MYANMAR, Mergui, S Tenasserim;

Tetragonilla collina (Smith, 1857)
   Trigona collina Smith 1857: Bingham 1897: 560, 562 (distribution, key to species, taxonomy); MALAYSIA, Malacca; MYANMAR, S Tenasserim;

Tetragonula iridipennis (Smith, 1854)
   Trigona iridipennis Smith 1854: Bingham 1897: 560, 563-564 (distribution, key to species, taxonomy); MYANMAR, Tenasserim; SRI LANKA;

Tetragonula laeviceps (Smith, 1857)
   Trigona læviceps Smith 1857: Bingham 1897: 560, 564 (distribution, key to species, taxonomy, uncertain identity (as praeterita)); MYANMAR, Rangoon; SRI LANKA, Trincomali; Bingham 1897: 560, 563 (distribution, key to species, nest (often build in crevices in the brickwork of the walls of houses), taxonomy); MYANMAR, Tenasserim; SINGAPORE;

Figure 79 Tetragonula laeviceps (Smith, 1857)

Tetragonula ruficornis (Smith in Horne & Smith, 1870)
   Melipona smithii Bingham 1897: 560, 563: Unnecessary replacement name for Trigona ruficornis Smith, nec Lamarck. Lamarck 1817 (repeated in 1835) only listed *Melipona favosa, M. amalthea, M. ruficrus, M. postica,* and *M. pallida.* The replacement name, therefore, remains enigmatic unless mistaken for *M. ruficrus* Latreille. BMNH has a labelled type for this replacement name (17b.1129) (distribution, key to species, taxonomy); INDIA, Mainpuri, NW provinces; MYANMAR, Tenasserim;

Tetrigona apicalis (Smith, 1857)
>*Trigona apicalis* Smith 1857: Bingham 1897: 560, 562 (common name (dammar bee), distribution, key to species, taxonomy (see also binghami)); MALAYSIA, Borneo; MYANMAR, Tenasserim;

Tetrigona binghami (Schwarz, 1937)
>*Trigona apicalis* variety *binghami* Schwarz 1937: 288, 300, 301, 303-304*, 328, plate 2, 5, 7: Holotype (BMNH 17b.1142); paratypes (AMNH, BMNH, USNM) (key to species, taxonomy); Type locality: MYANMAR "Tenasserim: Dawnat Range, Jan. 1891 (Col. C.T. Bingham)" (1 worker);

References for Vernacular architecture in Myanmar

Moe, S. S. & Hlaing, M. (2017). Study on Architectural Aspect of Traditional Bamar (Myanmar) Timber Houses. *International Journal for Innovative Research in Multidisciplinary Field* Volume - 3, Issue - 8-, ISSN – 2455-0620.

Oranratmanee, R. (2018). Vernacular Houses of The Shan in Myanmar in The Southeast Asian Context. *Vernacular Architecture*, Vol. 49 (2018), 99–120, DOI: 10.1080/03055477.2018.1524217.

Sriruksaa, K. & Tuansirib, P. (2020). Identity and the Wisdom of Vernacular Architecture in Kengtung, Shan State, Myanmar. *International Journal of Innovation, Creativity and Change.* www.ijicc.net Volume 14, Issue 1,

Zune, M . (2019). Vernacular Passive Design in Myanmar Housing for Thermal Comfort. Retrieved from Nottingham-repository. Worktribe: https://nottingham-repository.worktribe.com/preview/4829593/Manuscript_SCS_2019-1029_R2.pdf

Vernacular Passive Design in Myanmar Housing[73] for Thermal Comfort.

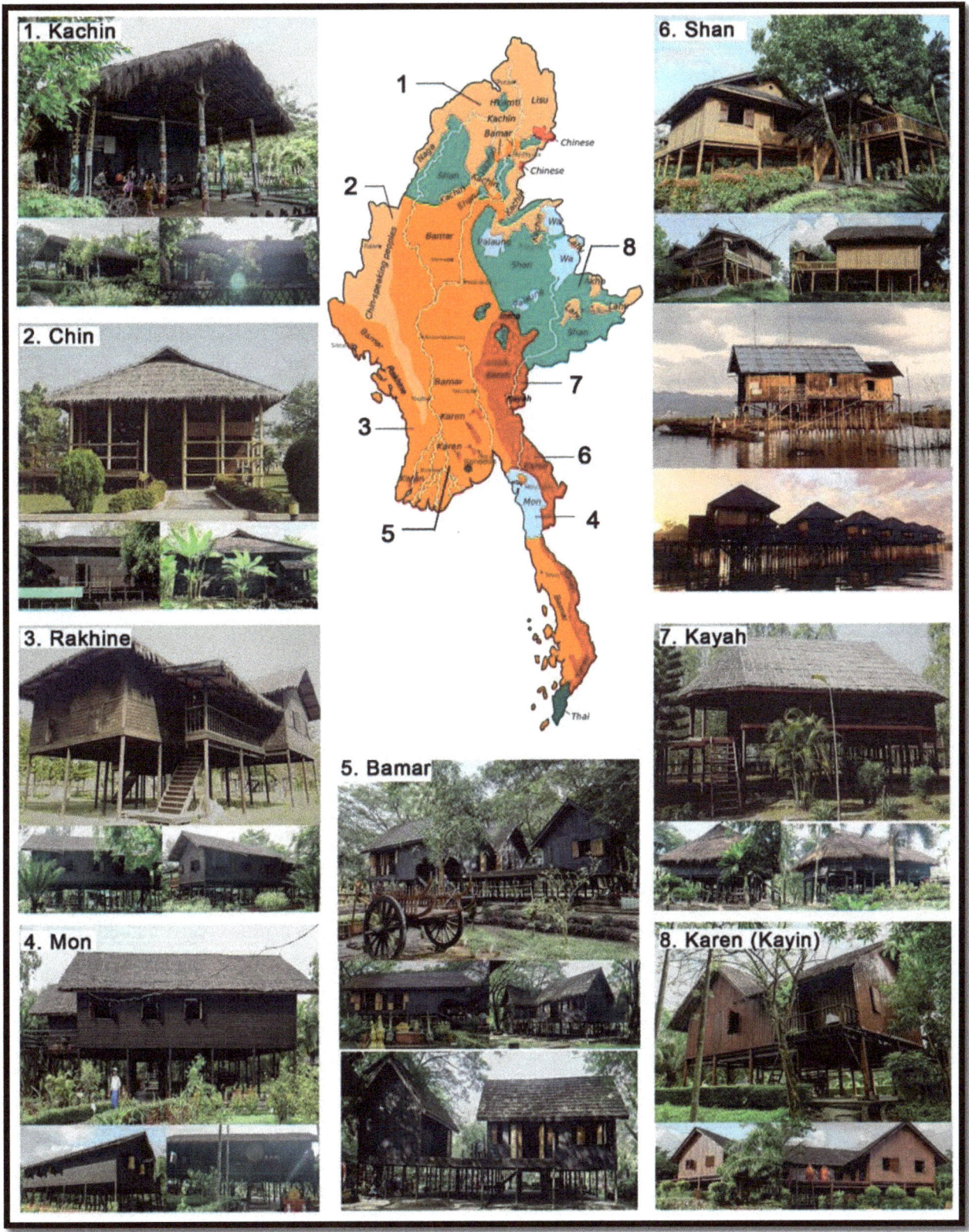

Figure 80 Regional Myanmar Housing

---

[73] Retrieved from Nottingham-repository. Work tribe: https://nottingham-repository.worktribe.com/preview/4829593/Manuscript_SCS_2019-1029_R2.pdf

Chapter 7

Stingless Beekeeping and Bee Plants in Thailand

By Anchalee Sawatthum (email address: anchsawat@hotmail.com)

Faculty of Agricultural Technology, Rajamangala University of Technology Thunyaburi

2 Paholyothin 87 Thunyaburi Pathumthani 12130

Introduction

Thai people in all parts of Thailand have known stingless bees as indigenous animals and have used stingless bee products for many purposes. Thailand is known as one region with abundant stingless bee species. Taxonomic studies confirm at least 22 species since 1854 in Thailand (Schwarz, 1939). Anuchit and Surapong reported an unrecorded species in Thailand in 2001. 2003 *Tetragonula sirindhornae* Michener and Boongird were reported as new species in Thailand (Boongird, 2003). Almost all stingless bee products were hunted from natural colonies.

Stingless bee culture

Stingless bee culture for specific purposes in Thailand was known in this decade. From the survey between 2000 – 2004, the formal stingless bee culture was found in two regions, namely the South and the East. In the South, a Muslim beekeeper kept *Tetragonula pagdeni* colonies around his house to harvest the honey.

This man began his beekeeping initially through observations. After the honey's initial harvest, he kept the stingless bee colony in bamboo stems or old betel palm trunks around 70-90 cm. long. He has seen the bees occupying sections of thick cylindrical shape bamboo commonly used for palm syrup harvesting.

The stingless bee-keeping procedure was as follows: The bamboo stem was split into two pieces (long section). All of the brood cells and storage cells from the old colony were placed on one piece of bamboo trunk and covered with the other, tied together with wire, closed the two open ends with coconut husk, and tied the trunk again with a long tie to hang it on the house's wall outside his house. The bees' food resources were around his house, such as coconut, betel palm, sugar apple, etc.

But stingless bee culture in the East was mainly found in the Chantaburi and Trat provinces. It was started by the Agriculture Occupation Development and Promotion Centre Chantaburi Province (Beekeeping) in 1999.

This centre has aimed to promote stingless bee culture in a responsible area (the East) to use the bees as insect pollinators of many fruit crops such as rambutan, litchi, cashew nut, etc. In the beginning, it was found that some farmers were keeping the stingless bee colonies at home. The farmer's houses in the East are normally in the orchards near the forest. Almost all bee colonies were moved from the forest into unused hollow containers such as big jars, tool boxes, and wooden boxes. Some farmers have collected the colonies found in decaying logs, live trunks, or branches in the forest. They placed the colonies under the eaves around their house or built small huts to keep them and protect them from the hot sun and heavy rain. In 2000, the centre promoted farmers to use insect pollinators to increase fruit yield.

Stingless bees are one of those which is easier to manage than honey bees. After the promotion, the demand for stingless bee colonies was very high. Farmers have brought many colonies cut from the living trees in the forest, especially from mangrove areas. In 2002, encroaching mangrove area was prohibited by the forester.

The centre officers have taught the farmer to box the nest since 2000. The original log (tree trunk or branch) must be split into two parts as gently as possible. The brood cells and storage cells were removed into a wooden box volume around $20 \times 30 \times 35$ cm^3 with a small hole on one wall near the base of the box as the nest entrance. Sticking resin from the original nest entrance onto the new box entrance will help the bees accept their new home. Place the box in the same position as the original log to encourage any flying bees to enter the new box.

The farmers in this region keep four species: *Tetragonula pagdeni* Schwarz, *T. laeviceps* Smith, *Lepidotrigona terminata* Smith and *L. ventralis* Schwarz. The dominant and most kept species is *T. pagdeni*. All species adapted themselves readily to the conditions of the wooden hive, and the farmers kept colonies successfully.

Differentiation of Stingless bee nests

The structure of the nest entrance, the aggressiveness of the guard bees and nest architecture are often useful for bee classification (Sakagami et al., 1990). In this paper, the characters of these four species are described in brief.

Tetragonula pagdeni Schwarz

The guard bees are relatively aggressive, and *T. pagdeni* has a cluster-type brood nest. The nest entrance of the *T. pagdeni* colony is tube-like, normally short with a dark colour or varies to the plant resin colour around the nest. In the artificial hive, storage cells are built near the entrance inside the hive or attached to the box's wall. The brood cells are built next to the food cells and can be attached to the box's wall, even on the cover.

Figure 81 Tetragonula pagdeni Schwarz

But in a natural colony in the log, brood cells are normally built in the middle of the hollow, whereas the storage cells are built near the entrance and the end of the hollow. It means that the brood cells are built in the middle between the storage cells and the hollow. It will be practical for the nurse bees to feed the food into the brood cells during cell building.

Figure 82 Top left: Nest entrance; Top Right: Internal nest structure; Bottom left and middle: Brood cells (Cluster–type nest) Bottom right: Storage cells

Tetragonula laeviceps Smith

The nest entrance of *Tetragonula laeviceps* is tube-like with a dark colour. The guard bees are aggressive. It has a cluster-type brood nest. The nest architecture in the box is similar to that of *T. pagdeni* but different in body size. Because the body size of *Tetragonula laeviceps* is relatively bigger than that of *T. pagdeni*, the brood cells and storage cells of *T. laeviceps* turn out to be bigger than those of *T. pagdeni*.

Figure 83 Tetragonula laeviceps Smith

Figure 84 Left: Nest entrance; Middle: Brood cells; Right: Storage cells

Lepidotrigona terminata Smith

L. terminata has a flare(bell–mouthed) shape funnel with light-coloured material. The guard bees are not aggressive. It has a horizontal spiral brood comb. The brood comb is built with advancing front spirals upwards from the bottom of the box. The storage cells are built and attached to all walls of the hive. There is an involucrum between the brood cells and storage cells. The honey pots and pollen pots are bigger than those of *T. pagdeni*.

Figure 85 Lepidotrigona terminata Smith & Nest entrance

Figure 86 Left: Brood cells (Horizontal spiral brood comb); Middle: Storage cells; Right: Involucrum

Lepidotrigona ventralis Schwarz

Its nest entrance and nest architecture are similar to those of *L. terminata*. The behaviour of the guard bee is also similar.

Figure 87 Lepidotrigona ventralis Schwarz & Nest entrance

Figure 88 Left: Brood cells; Middle: Involucrum; Right: Storage cells

Natural enemies

It was found that some predators and parasites were attacking the stingless bee colonies. Predators were the *Vespa* hornet and soldier flies (Stratiomyidae). The *Vespa* hornet was found to attack the foraging bees in front of the hive during flight, whereas reduviid bugs stayed and caught the adult bees near the hive entrance. Soldier flies attacked the weak colony by eating the brood and storage cells. Stingless bee parasite was found only in the *T. pagdeni* colony, a parasitic mite that attacked the brood cells.

Utilization of stingless bees

Honey harvesting

The Muslim Beekeeper has only harvested honey from *T. pagdeni* colonies and sold it commercially in the south. He has more than twenty colonies hung around his house. Each colony was harvested once or twice a year. He has filled honey in the bottle (750 cc) and sold it for 14 US Dollars, whereas the honey from *Apis mellifera* in the same volume is sold for 5-10 US Dollars. Nowadays, the farmers in the East also sell stingless bee honey, a bottle for 40-60 US dollars.

Use as an Insect pollinator.

The stingless bees bear a high potential to pollinate wild and domestically cultivated plants. In eastern Thailand, the farmers have used the stingless bee as an insect pollinator for their fruit crops, such as rambutan and litchi.

To induce the potential of stingless bees for pollination, some farmers have placed the colonies in their plantations, but others have only left them around their houses.

Stingless bee plants

Food plants are important in stingless bees' survival or colony development. Stingless bees visit diverse plant species of flowers to collect pollen grains and nectar. The study of stingless bee pollen plants was conducted in two apiaries in the experimental farm of the Faculty of Agricultural Technology in Pathumthani Province (near Bangkok) and the stingless bee farm in Chantaburi Province (Eastern part of Thailand). The observations were conducted during the wet and dry seasons. Stingless bee pollen plants were confirmed by the flowers' visitation of stingless bees and pollen analysis from pollen pellets collected from the stingless bee in front of the bee hive. Pollen sources plants of stingless bee from observation results could be divided into three groups as follow:

1. Ornamental Plants

There were eight species of ornamental plants found. These were Cosmos (*Cosmos* spp.), Rose (*Rosa* hybrids), Rose moss (*Portulaca oleracea* L.), Water Lily (*Nymphaea lotus* L.), Coral Vine (*Antigonon leptopus* Hook and Arn.), Yellow flame (*Peltophorum pterocarpum* Baacker ex K.Heyne), Sacred Lotus (*Nelumbo nucifera*), and Angsana (*Pterocarpus indicus* Willd).

2. Vegetable Plants

Many species of growing plants used as vegetables were found as pollen sources of stingless bees, such as Banana (*Musa sapientum* L.), Bird Chilli (*Capsicum frutescens* Linn.), Hairy Basil (*Ocimum basilicum* Lif. var. *citratum* Back.), White gourd (*Benincasa hispida*), Holy basil (*Ocimum sanctum* L.), White popinac (*Leucaena leucocephala* (Lamk.)), Vegetable Humming Bird (*Sesbania grandiflora*), Moonflower (*Ipomoea alba* L.), Corn (*Zea mays*), Chineses spinach (*Amaranthus lividus* Linn.) and Sunflower (*Helianthus annuus* L.).

3. Weeds

Only two weed species were found and can be confirmed as pollen plants of stingless bees. These were 'Cat's Whiskers' (*Orthosiphon aristatus* (Blume)) and Margarita (*Bidens pilosa* var. *radiata* (L.) DC.). The observation's results showed less diversity of plant species that can be the pollen source of a stingless bee because this study was based on preliminary observations in a limited area. More observations in larger areas and all year-round observations are needed to get more details about stingless bee plants.

Conclusion

Stingless bee diversity in Thailand is risky because of forest encroachment and a lack of basic knowledge about bees.

Concerning this study, the stingless bee culture in Thailand has just started and still needs more research to help the stingless beekeepers manage the colony for specific purposes such as honey harvesting and use as an insect pollinator.

Acknowledgements

I would like to give my sincere thanks to Prof. Dr. Michener for his identifying the stingless bee species. I would also like to thank the Agriculture Occupation Development and Promotion Center officers and Mr. Suchat Tulkul for their valuable information. The Rajamagala University of Technology Thunyaburi for a research grant

References

Anuchitchinajariyawong and Surapongsaibun. 2001. *Trigona laeviceps* Smith culture, p.445-449 *In* Proceeding of 38th Kasetsartuniversity Conference Bangkok.

Boongird, S. 1992. Biological studies of stingless bee, *Trigona laeviceps* Smith and its role in the pollination of durian, *Durio zibethinus* L. cultivar change. PhD thesis, Kasetsart University, Bangkok.

Boongird, S. 2003. *Trigona sirinclhornae* Michener &Boongird. Ramkamhang news 33 (27) : 5

Sakagami, S.F., T. Inoue and S. Salmah. 1990. Stingless bee in central Sumatra. pp.125-138. *In* S. F. Sakagami, R. Ohgushi and D.W. Roubik (eds), Natural History of Social wasps and Bees in Equatorial Sumatra. Hokkaido University press, Sapporo

Schwarz, H.F. 1939. The Indo – Malayan species of *Trigona*. Bulletin of the American Museum of Natural History LXXVI, part III.pp.83 - 141

Chapter 8

Vernacular Architecture of Thailand

Thai houses of the four regions

The houses from the four regions may all look alike at first glance. But there are distinguishing features in each type of house. The differences are in the gable roof the triangular side of the house between the sloping sides of the roof.

Figure 4 Houses from the Northeast region or Issan have a "Kaan" (bamboo organ).

Figure 4 Northern houses have a "kalae" (V-shaped carving at the gable peak)

Figure 4 Houses in the Central region have horn-like structures at the base of the gable

Figure 4 Southern houses have an ornate carving around the rod at the apex of the gable

Thai Houses

Like other houses in Southeast Asia, the Thai house is a wooden structure raised on posts. Over many centuries, it has acquired its unique style. The distinguishing marks are an elegantly tapering roof and various finials and decorations that differ regionally. While architectural features vary throughout the four cultural regions, Central Thailand, the North (*Lanna*), the North-East (*Isaan*), and the South, raising a platform on poles is common to all parts of the country. It offers protection from dirt, hostile wildlife, thieves, and, most importantly, from the monsoon floods that affect all of Thailand.

The traditional Thai house is ideally adapted to its environment. The high-pitched open roof facilitates air circulation. Open windows and walls and a large central terrace provide ideal ventilation and relief from the hot and humid climate. Wide overhanging eaves protect the house from sun and rain. Rainwater quickly runs off the steep roof and falls through the permeable terrace and house floors. The use of wood and bamboo reflects the once-abundant forests that provided these materials ubiquitously and cheaply. In the past, an agricultural society existed in relative harmony with its natural environment.

One may hope that the tropical climate will do its part to rid the landscape of unsightly and poorly adapted structures and that the commencing rediscovery of vernacular architecture will lead to increased harmony between buildings and the environment.

Traditional And Contemporary Houses in Thailand

Over the years, people in Thailand have adapted to the changes brought about by technological developments. However, modern concrete structures in Thailand reflect little of the traditional wisdom inherited from the old design. The traditional house today cannot serve the variety of activities in the same way as the contemporary house, which is cheaper, stronger and more flexible and therefore appears more popular than the traditional house.

Design layout and materials

Figure 5 Thai Rake Boards

Figure 6 Typical Thai Finial

Some contemporary houses, shown in Figure 94, are constructed with concrete structures and floors, brick walls, and a plasterboard ceiling with 3cm of insulation and concrete tiles on the roof. The windows are single-glazed. This feature is typical of houses currently being built in all regions of Thailand. The traditional Thai house is ideally adapted to its environment. A high-pitched open roof facilitates air circulation. Open windows and walls and a large central terrace provide ideal ventilation and relief from the hot and humid climate. Wide overhanging eaves protect the house from sun and rain. Rainwater quickly runs off the steep roof and falls through the permeable terrace and house floors. (Text by Thomas Knierim)

Design for a climatic responsive contemporary house in Thailand[1]

The question posed in the paper was whether traditional Thai houses perform better than the typical contemporary Thai house in creating comfortable internal conditions. Therefore, the thermal performance of both types of buildings is investigated using a simulation model based on a selection of thermal performance criteria of local traditional and contemporary house models in Thailand. The results indicate that improved performance might be achieved by combining selected lessons from the

---

[1] Lessons from traditional architecture: Design for a climatic responsive contemporary house in Thailand
Paruj Antarikananda, E. Douvlou, K. McCartney - Published 2006 Semantic Scholar Corpus ID: 45053458

traditional design, e.g. improved shading, regional variations in window size related to orientation, and adoption of adjustable ventilation and window openings.

Comfort Indicators: The cooling loads are significantly higher than the heating loads, except for the traditional house in the south zone, where both heating and cooling loads are relatively small. Therefore, the response to the threat of overheating was taken as the primary indicator of the success of the house design in responding to climate.

Figure 7 Regional Types of Thai House (Redrawn from Antarikananda, 2005)

Three indicators of response to overheating are used:
(1) duration of overheating,
(2) intensity of overheating, measured in degree-hours/year and
(3) cooling load, which is an estimate of the useful energy required to keep the internal temperatures within the comfort zone.

The traditional houses selected for comparison with the contemporary houses in the south, central, and northeast regions of the country are illustrated in Figure 92. They are constructed almost entirely of wood, with no ceilings and clay-tiled roves. There is no glass in the window apertures. A detailed description is available.

The paper concluded that Thailand's traditional housing provides useful indicators of appropriate architectural design responses to climate, particularly in the context of purely passive environmental control. The traditional house designs are superior to the contemporary in providing thermal comfort for Thailand's three selected climatic zones. The design of the contemporary house may and should be informed by that of the traditional house. However, issues of lifestyle requirements and cultural issues should also be considered in the final design.

Stringless bees with Type locality records in Thailand:

Homotrigona aliceae (Cockerell, 1929)
Trigona aliceæ Cockerell 1929b: 139, 140: Holotype (BMNH 17b.1116) (comparative note, taxonomy); **Type locality**: THAILAND "Siam: On a trail in the jungle between Pahtoop mountain and Nan, Jan. 9, 1928(Alice Mackie)" (worker);

Homotrigona fimbriata (Smith, 1857)
Melipona castanea Bingham 1903: vi: Type (BMNH 17b.1128) (taxonomy); **Type locality**: THAILAND "Bukit Besar, Nawngchik. 1500 to 2500 feet" (worker);

Homotrigona lutea (Bingham, 1897)
Trigona ferrea Cockerell 1929b: 139-140: Holotype (BMNH 17b.1117) (comparative note, taxonomy); **Type locality**: THAILAND "Siam: Mekami River, Feb. 3, 1928 (Cockerell)" (worker);

Lepidotrigona doipaensis (Schwarz, 1939)
Trigona (Lepidotrigona) ventralis variety *doipaensis* Schwarz 1939a: 85, 94, 136*: Holotype (USNM 53563, worker): examined, "Doi Pa / Mai Deng / Siam 750m / 12-29-32", "Hugh Smith / Coll", "Type No./ 53563 / U.S.N.M.", "Holotype", "Trigona ventralis / var. doipaensis / H.F. Schwarz". The ventralis species group (distribution, key to species, taxonomy); **Type locality**: THAILAND "SIAM. -Doi Pa, Mai Deng, 750 meters, Dec. 29, 1932 (H. M. Smith)" (worker);

Lepidotrigona flavibasis (Cockerell, 1929)
Trigona flavibasis Cockerell 1929c: 592: Holotype (AMNH). The ventralis species group (taxonomy); **Type locality**: THAILAND "Doi Sutep, Siam, Feb. 9 (Alice Mackie)" (worker);

Lisotrigona furva Engel, 2000
Lisotrigona furva Engel 2000: 235-236: Holotype (SEMC); 5 paratypes (AMNH, SEMC) (floral record, key to species, taxonomy); Type locality: THAILAND "THAILAND: Nakhon Ratchasima Prov.: Tha Chang, 10 February 1993, S. Boongird and C. Michener" (worker); THAILAND, Chiang Mai, W Fang; Floral record: *Buddleja asiatica;* Callistemon;

Tetragonula hirashimai (Sakagami, 1978)
Trigona (Tetragonula) hirashimai Sakagami 1978: 166-194, 195, 196, 197, 198-200*, 202, 212, 215, 226, 227, 229, 234, 236, 237, 238, 239, 240: Holotype (SEHU) (taxonomy); **Type locality**: THAILAND "Chieng Dao-b, N. Thailand" (1 worker, holotype); "Doi-Suthep" (8 workers, 62 males)"; "Chieng Mai (1 worker); "Mae Klang" (2 workers); "Mae Fack" (1 worker); "Lamphun" (1 worker); "Uthaithani" (1 worker);

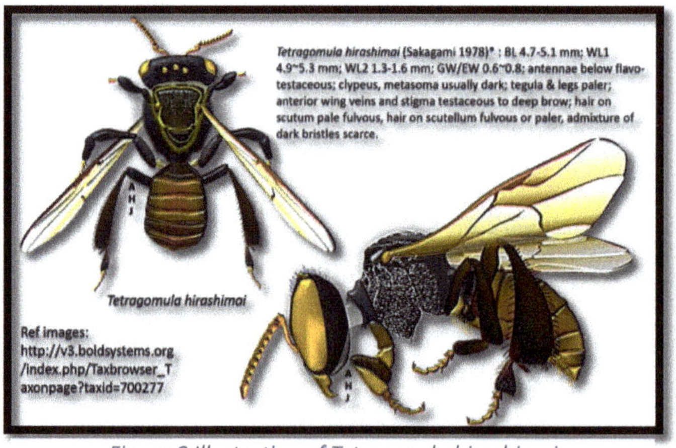

Figure 8 Illustration of Tetragonula hirashimai

Tetragonula pagdeni (Schwarz, 1939)
Trigona (Tetragona) fusco-balteata variety *pagdeni* Schwarz 1939a: 85, 93, 96, 110-111*, 113: Holotype (USNM 53564, male): examined, "Nakon / SriTamarat / Siam 7/5/28", "Hugh Smith/coll", Type No. / 53564 / U.S.N.M.", "Holotype", "Trigona / fusco-balteata / var. pagdeni / H.F. Schwarz"; paratype (AMNH (1)) (distribution, key to species, taxonomy); **Type locality:** THAILAND "SIAM. -Nakon Sri Tamarat, July 5, 1928 (Hugh M. Smith)" (workers, males); "Singora, June 1929 (Hugh M. Smith)" (workers);

Tetragonula pagdeniformis (Sakagami, 1978)
Trigona (Tetragonula) pagdeniformis Sakagami 1978: 166-194, 209, 211-213*, 226, 227, 229, 234, 236, 237, 238, 240: Holotype (EKYU, not located, nor in SEHU); paratypes (BMNH (5)) (taxonomy); **Type locality**: THAILAND "Khao Chong-2, Peninsular Thailand" (1 worker, holotype);

Tetragonula sirindhornae (Michener & Boongird, 2004)
Trigona (Tetragonula) sirindhornae Michener & Boongird 2004: 143-146: Holotype (SEMC, worker); paratypes (AMNH (3), Insect Museum, Chakthong Building, Department of Agriculture, Bangkok, Thailand (3), SEMK (11) (taxonomy); **Type locality**: THAILAND "Thailand, on the Isthmus of Kra: Ranong Province: Muang district, 9–13 November 2002" (11 workers); "Ngow Waterfall National Park" (7 workers);

Tetrigona apicalis (Smith, 1857)
Syn. Trigona hemileuca Cockerell 1929b: 140: Holotype (BMNH 17b.1098) (comparative note, taxonomy); **Type locality**: THAILAND "Siam: Nan, Dec. 31 and Jan. 31 (Cockerell)" (worker);

Tetrigona melanoleuca (Cockerell, 1929)
Trigona melanoleuca Cockerell 1929b: 140-141: Holotype (BMNH 17b.1119). Possibly a junior synonym of vidua. See below (comparative note, taxonomy); **Type locality**: THAILAND "Siam: Nan, Jan. 13, 1928 (Alice Mackie)" (worker);

Figure 9 **Tetragonula (Tetragonula) malaipanae** Engel, Michener & Boontop 2017

Tetragonula (Tetragonula) malaipanae Engel, Michener & Boontop 2017 [Etym: The specific epithet honours Savitree Malaipan of the Department of Agriculture, Kasetsart University, Bangkok, beloved mentor of Y.B. and a prominent Thai melittologist.] Holotype: Thailand, Kanchanaburi [Province], Thong Pha Phum [District], 17. ix.2007 [17 September 2007],

Lepidotrigona satun Attasopa and Bänziger n. sp. 2018 Etymology. The specific epithet refers to the province in Thailand where the species was collected;

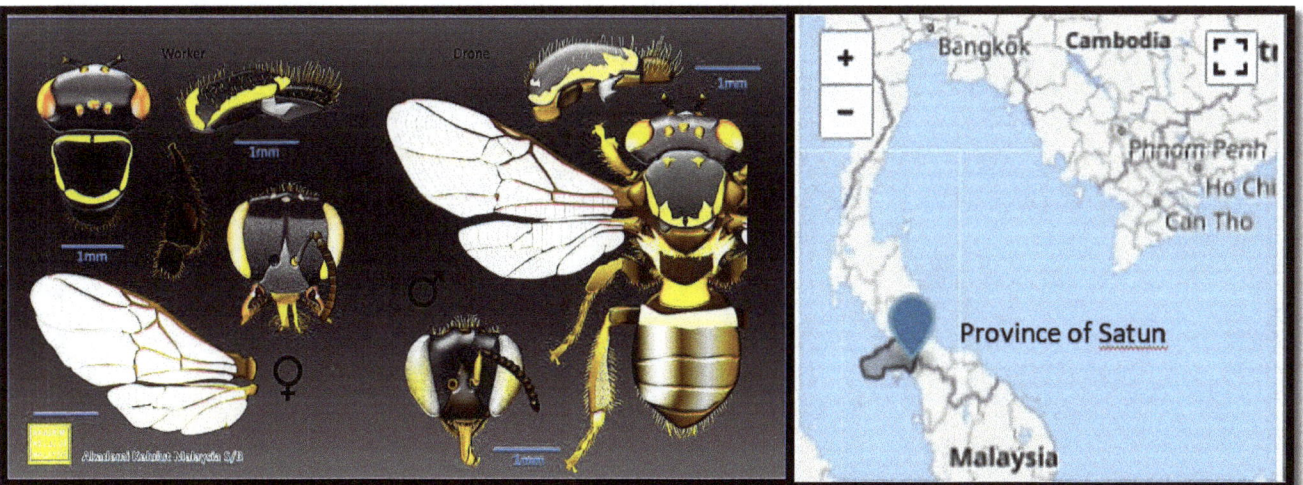

Figure 10 Lepidotrigona satun collected in the Province of Sarun near the Thailand – Malaysia border

Bamboo huts for the bee farm.

The bamboo hut is considered another identity of the residence of the Thai people. Since ancient times, the way of life of most people in the country has been associated with a farming occupation in some

Figure 11 Assorted Bamboo huts for Bee friendly environment

form or another. Therefore, we are often familiar with small bamboo huts, which are situated near the edges of the fields (for ruminants) and crops. Bamboo houses or bamboo huts of each locality have different beauty and charm.

Below are ideas for 15 different bamboo huts. They can be used as shelters on the head of the bee farm, crop field or backyard. The bamboo hut is designed with premium workmanship, perfectly retaining the Thai way of life and style. It can be built as a backyard shelter or rest hut. Some are simply beautiful and do not break any particular rule. Some are like a family-style bamboo cottage house, cooling and cosy like any cottage can be. Put it in the garden, and it will enhance the overall aesthetics.

Wall cladding

Differences in the design depend on a family's financial means and the materials available. Houses of simple farmers generally have walls made of palm leaf matting, the preparation of which is labour-intensive but does not rely on imported materials. The palm leaf matting is fixed directly to the structural framework. Fine bamboo struts are often used to anchor the matting. In more sophisticated

Figure 12 Types of wall cladding

houses, wooden boards are used to clad the walls, aligned either horizontally or vertically. The horizontal version is fixed like weatherboarding, with the upper board overlapping the board below. Boards are butt jointed; sometimes, joints are covered with a bamboo strip.

We observed the unique Thai vernacular architecture when visiting Dr. Bänziger in Chiang Mai. Dr. Bajaree Chuttong was there with Sangdao Bänziger, Dr. Hans Bänziger, me and my wife, Zaiton. We were having dinner in a famous Muslim Restaurant. It is fairly easy to find Mosques or Muslim food in Thailand. There are many other architectures, cultures and faiths you may find there.

Figure 13 Dinner in Muslim Restaurant in Chiang Mai. From left: Dr. Bajaree Chuttong, Sangdao Banziger, Dr. Hans Banziger, AHJ and AHJ's wife Zaiton.

Laos Traditional Housing

Figure 14 The Akha Tribal Hill slope House and a traditional house in Namo Nua Village, Laos.

Kretz, E. 2009 has researched traditional housing in northern Laos – wood preferences and impact on forest diversity. The research selected two villages (Namo Nua and Phou Xan) with different ethnicity and wealth standards in northern Laos.

Figure 15 A floating tomato garden on Inle Lake

According to the research, construction skills and knowledge about timber are given the utmost importance in the village of Namo Nua. Namo Nua villagers build their houses (Figure 98) on high poles, several meters over the ground, and use varied tree species in their constructions. Leaves from rattan or grass are used as a roof. As per the research, the largest houses comprised roughly one tree for the walls and one for the floor. Only a few nails are used in the construction, as it is built to be fixed together like a puzzle.

Chapter 9

Kampuchea Vernacular Architecture

The stingless bee of namesake significance in Kampuchea is *Tetragonilla **cambodiensis*** (Cockerell 1926) [*Typ. Loc.:* **Cambodia** "Angkov (=Angkor) Wat, Cambodia, Jan. 11, 1926 (H.M. Smith)] (Rasmussen, 2008) = Pertaining to Cambodia. This species epithet is a synonym for *Tetragonilla collina* (Smith, 1857)[75] (Rasmussen et al. 2017). *T. collina* is a common stingless bee species in Indo-China and Malaysia. They frequently nest in buttress roots and build foundation crevices.

Cambodia has changed its name several times. Between 1953 and 1970, the country was renamed the Kingdom of Cambodia and then the Khmer Republic until 1975. Under communist rule from 1975 to 1979, it was referred to as Democratic Kampuchea.

Exploring the traditional architecture in Kampuchea, this chapter will examine the domicile structures of the indigenous Khmer and the Cham people. The Cham people are from the Kingdom of Champa[76], once part of the Khmer Empire.

How many styles of Khmer traditional houses are there?

Chea Samath, professor of Architecture at the Royal University of Fine Arts in Phnom Penh, said that "the architectural styles of Khmer houses followed **five main types** [Phteah-Khmer, Phteah-Pit, Phteah-Kataing, Phteah-Ruang-Dual, and Phteah-Ruang-Doeung].

Historical Proofs of the Funan[77] Era

According to the book Cambodia's Funan Era, written by Dr. Michael Trané, a former secretary of state for the Ministry of Culture and Fine Arts, archaeological diggings have yielded evidence showing Funan as Cambodia's first kingdom between the first to sixth centuries AD.

Homes built by the common people during this ancient era were made of wood and leaves, covering the stilts erected to raise the houses. According to an ancient Chinese record written by a Chinese tradesman who came to Cambodia during the Funan era, these houses traditionally featured thatched roofs. On the other hand, the homes of the wealthy were normally covered with tiles made of baked clay.

---

[75] Source: https://www.itis.gov/servlet/SingleRpt/SingleRpt?search_topic=TSN&search_value=1129565#null
[76] the Kingdom of Champa is now South Vietnam.
[77] Funan was the first large Southeast Asian civilization. It was centred on the lower Mekong Delta in present-day Cambodia and Vietnam and stretched into Thailand, and, possibly the Malay Peninsula. (https://factsanddetails.com/southeast-asia/Vietnam/sub5_9a/entry-6637.htm)

An academic paper by Srey Ou of the Conferring Team of Khmer Traditions focuses on the periods from the Angkor era to King Sovanakoad, wherein favourite styles that prevailed from the Funan era were Phteas Rong Doal, Phteas Koeng and Phteas Khmer. These homes were built to align with the sunrise[78] and sunset and faced east to respect the Hindu Sun God – who the ancient Khmers believed to be Preah Aditya, son of Goddess Atite.

During the Angkor era, Zhou Daguan (or Chou Ta-Kuan), China's diplomatic ambassador to the Khmer empire of the time, noticed and recorded his observation of the Cambodian people living on stilt houses constructed with wood, thatch, and clay-backed roofs. The king's palace and homes of royal families and the wealthy were built the same way, except they were decorated with gold and other precious metals.

The Angkor Borei Archaeological Museum in Takeo province houses physical proofs of materials used to build the houses of the Funan era. These materials include large baked clay pieces and soil-hardened bricks.

Khmer House for Ordinary People

1. Pteas Pit-- It has four small roofs combined.
2. Pteas Rongdorl has open spaces beneath the roof to provide natural ventilation.
3. Pteas Rongdeung has a big roof and a pedestal near the front.
4. Pteas Kontaing — The roof is a triangular portion of roof pitches facing each other.
5. Pteas Khmer — It has two roofs, making a sloping slope.
6. Pteas Koeng has two rooftops overlaid with two heads in front and rear. It is also built for senior officials.

Traditional Khmer Housing[79]

It refers to the construction and assigned usage of houses or buildings by the Khmer people since ancient times and has evolved until today. In Cambodia, many Khmer-style houses are built differently depending on hierarchy and purposes. In special terms, the house symbolises prosperity in the national society and serves the people's lives in each village, which is culture and nature. The Khmer have long been known to live in different designs of stilt houses traditionally, with multi-levelled floors and gable finials at both ends of the ridge.

---

[78] Stingless bees have a tendency to face their exit/entrance tubes facing the sunrise because that is the earliest that the common nocturnal predators like frogs and lizards would retreat by the break of dawn.
[79] Source: https://en.wikipedia.org/wiki/Traditional_Khmer_Housing

History

An image of the Khmer traditional house in front of Angkor Wat was taken between 1919 and 1926.

Houses and settlements

The earliest record of Cambodian housing is from the Chinese record of the Funan kingdom (1st-7th century AD), where the residents were described as people who

Figure 104 Khmer traditional house in front of Angkor Wat taken between 1919 - 1926.

lived on stilt houses, cultivated rice and sent tributes of gold, silver, ivory and exotic animals. The Chinese accounts also mentioned the walled towns, dwelling houses and palaces. The people built their houses over the ground and made them accessible by ladders, as seen across Cambodia today.

A wooden structure as depicted on 12th century-Bayon temple's relief, which shared a similar triangular roof concept as the roof of today's Khmer houses and pagodas.

During the Khmer Empire (8th-15th century), high-status people were known to live in large houses, parts

Figure 105 12th century-Bayon temple's bas relief

of which were covered in roof tiles, and commoners lived in smaller houses with thatched roofs from perishable materials to this day. These natural fibres are not preserved and are replaced periodically. Also, houses were built on stilts, so the living floor was above the ground, according to Zhou Daguan, who visited Angkor in the 13th century. Bas reliefs from the Bayon temple depicted houses, buildings, and palaces that shared similar roof designs and concepts with today's Khmer traditional houses and palaces.

The architecture resembles the Thai Buddhist influence as we enter Laos and Cambodia. However, the original *chofah* upon which most subsequent *chofahs* have been based on the *gajashimha* of Suryavarman II, the Khmer king who built Angkor Wat. These finials (*chofah*) symbolised the unification of the northern and southern Khmer kingdoms and the reign of King Suryavarman II. This symbolism spread throughout the region, including Laos, Lanna (Northern Thai), and Isan (NE Thai), once the Khmer Empire.

Chapter 10

Types of Khmer House for Ordinary People

A compound of traditional Khmer houses constructed along and over the pond resembling an elite house in the Cambodian Cultural Village, Siem Reap.

Khmer people construct their houses differently based on their social status, wealth, preference, and geographical location. However, what they share in common is they are usually built as stilt houses, where houses are raised on stilts over the surface of the soil or a body of water to prevent flooding. Floating houses were also built around the Great Lake region of Tonle Sap.

Pteas Pit

Depending on location, the Pteas Pit house is called "Pteas Jorm" or "Pteas Krorjorm". This type of house is based on the roof's structure, how it is made or how it looks. This house is built with an equilateral quadrangular roof without a fronted porch. Usually, looking at the house's layout, we can see that the door is only in the middle of the wall, which is the largest.

Figure 106 Pteas Pit house. Source: https://www.facebook.com/ASEANCambodia/photos

Pteas Rongdorl

Figure 107 Pteas Rongdorl

Rongdorl – Poetically depicts a curtain in the home of the "Rong". The house has four columns of pillars and two rows of pillar lint between the columns and the right pillar. The crown is built using three rows of columns. The highest central rows are called rostrum pillars; others are called piers. These middle column pins are larger than the side columns, with the rostrum function. In addition, it is often seen that the middle column pillar is elaborate.

Pteas Rongdeung

The house was constructed with a big roof and a porch at the front and rear before King Suriname's reign and King Ponhea Khemara's reign. Observations that tried to name the different types of Khmer houses are usually determined by the roof. The two-apartment home is overlaid with two roofs in front and back. The central pillars have two rows of rooftops. The connecting hallway can form a corridor if the second column covers it. The

Figure 108 Pteas Rongdeung or similarly a Cham House

corrugated plate on the corridor can be brought slightly lower than the width of the house. At times, the entrance door is made in four directions, and stairs at any door or in all directions. This depends on whether a family event demands the stairs at the width or the length of the house. For example, the stairs are usually at the centre of the house, but at times, such as at weddings or formal ceremonies, the children move the stairs to the side, not far from the ceremony.

Pteas Kontaing

The simplest Pteas Kontaing's roof is generally a triangular portion of roof pitches facing each other. The pillar has at least three rows. The highest centre row is called a rope, pole or column. Other pillar rows are called "*veang* rows". Normally, the veang rows are at the corner of the house. The front door is always at the top of the stairs. The structure is simpler than other house structures, only having large logs and some simple materials to build the house. Because it is common and beautiful, it is easily found in Cambodia, especially nowadays. Houses made of leaves can also be called Pteas

Figure 109 The simplest Pteas Kontaing's roof

Kontaing and are easy to repair.

Pteas Khmer

A type of traditional Khmer house is known as Pteas Khmer in classification.

Figure 110 *A type of traditional Khmer house known as Pteas Khmer*

Some Khmer houses have a high roof, and some don't employ Rongdorl or Rongdeung. Pteas Khmer houses have two roofs, making an inclined slope. A painting in the early 20th century shows one single home can be alone or in consecutive twin rows. Also, most of the houses are rented by monks when they see some rows drop down the roof, sometimes in front of the side, to create the next apartment. To make Khmer houses, they must have piles of long sticks for rooftops and use wood slats or planks. Quite often, people had the resources to afford the Khmer house, and it was deduced that they even built this type of house for the nobility during the reign of King Kulondak (Preah Thaong).

Pteas Koeng

Figure 111 *A type of Cambodian house that was built only for senior officials.*

A Cambodian house built only for senior officials, the initial base house was marked with two rooftops overlaid with two heads on both sides. From this base, it is further developed into other forms. The central pillars have two rows, and the intermediate connection can form a corridor if it is covered. A pitched roof plate on the corridor can be built slightly lower than the width of the house. This is a lively house built in 1907.

Traditional Rest House[80]

A rest house along a riverside was built in traditional Khmer architecture in the 1930s.

The construction of rest houses was famous during the reign of Jayavarman VII, the great king of the Khmer Empire. During his prosperous reign, Jayavarman VII built 121 "houses with fire" rest houses built every fifteen kilometres along raised highways for travellers, and 102 hospitals throughout his empire connecting Angkor to other Khmer cities such as Phimai (present-day Thailand), Vat Phu (present-day Laos), and Champa (present-day Central Vietnam). This tradition has been practised until today throughout Cambodia, where traditional wooden houses were built along the roads for the travellers to take a rest or shelter known as *sala samnak* (Khmer: សាលាសំណាក់).

Figure 112 Traditional Rest House Source: https://en.wikipedia.org/wiki/Traditional_Khmer_Housing#/media/File:Khmer_traditional_resthouse_in_1930.jpg

Traditional rural Khmer house[81]

Figure 113 A traditional Khmer house in the mid-1800s taken by Emile Gsell.

---

[80] From historical depictions Figure 171 Source:
https://en.wikipedia.org/wiki/Traditional_Khmer_Housing#/media/File:Khmer_traditional_resthouse_in_1930.jpg
They would make great Meliponary structures and excellent for bee farms as well as attractions in tourist parks.
[81] https://en.wikipedia.org/wiki/Traditional_Khmer_Housing

The typical rural house is the basic structure from which all variations have developed. Construction elements are of wood; rounded posts are used for piles, whilst joists are made of squared beams. Simple carpentry is used to join horizontal and vertical elements; the use of screws, which would increase expense, is avoided. Panels of homemade palm matting cover the sides of the house; fixed to the wooden structure, they merely protect against the elements but do not influence the stability of the construction. Palm matting is also used for roofing.

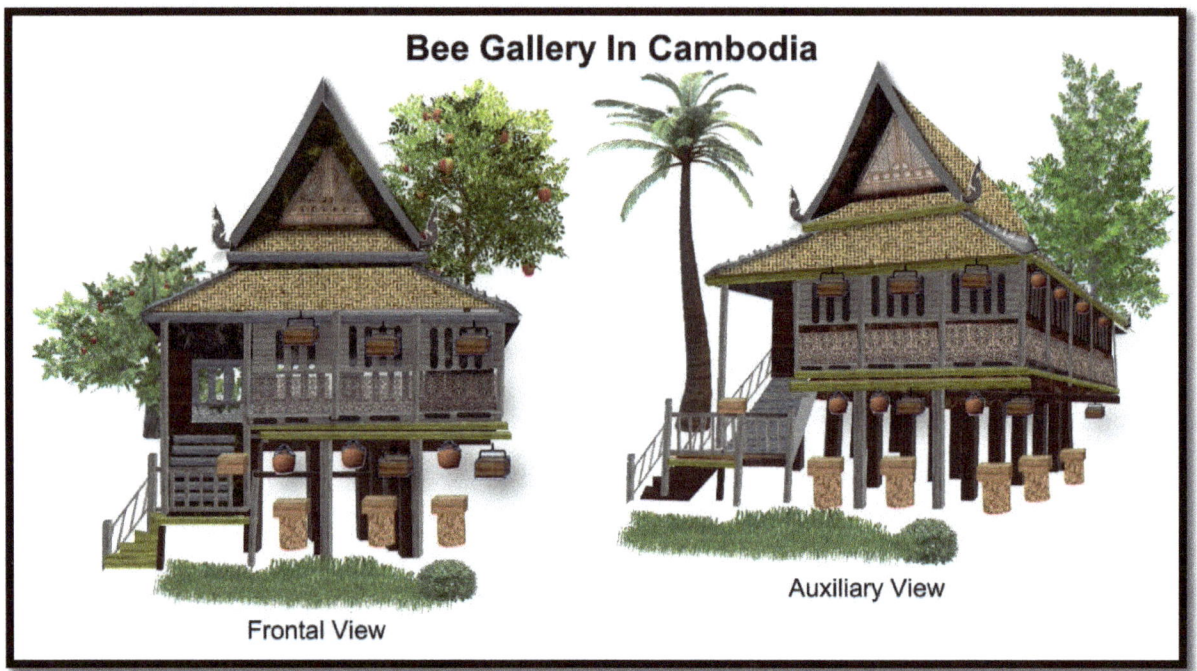

Figure 114 Bee Gallery in Cambodia – Basic roof design is Pteas Khmer

The Bee Gallery

This Bee Gallery (Figure 110) is styled from the Pteas Khmer house with ample headroom to stand under the house. The vertical slot windows are ideal for unrestricted airflow and incorporate a veranda with an ornate filigree guard rail or Indochinese carvings for tourist appreciation. Inside are wide spaces for racks to display assorted bee products.

Chapter 11

Traditional Cambodian houses

Model Houses are redrawn, and historic text excerpts from https://www.salalodges.com/.

Figure 115 Hobby Horse House (Phteah Chiahseh)

The "Hobby Horse House" (*Phteah Chiahseh*) takes its name from the two handrails, which end with a horse's head carved out of wood. This house was located in Kompong Chèn, in Kompong Chèn Tbon. This village, the "port of the southern Chinese, " is situated in the province of Kompong Thom, 80 kilometres from Siem Reap. This construction dates from 1956.

The "House with Grey Shutters" (*Phteah Betotvear Pnrbrapheh*), built in 1983, this house is from Wat Svay, a village in Salakamroeuk. It was owned by Dara Sin, a 49-year-old soldier and his wife, Kimouch Seng, a gold retailer. It was reconstructed on the site in June 2012.

Figure 116 House with Grey Shutters (Phteah Betotvear Pnrbrapheh)

The "House of Many Roofs" (*Phteah Dambaul Chraen*) distinguishes itself by its unique triple roof and horizontal cladding.

This house was spotted in Ta Prom, a village in Siem Reap province situated some 60 km. from the town. Built by a master carpenter in 1957, it is embellished with many decorative features.

Figure 117 House of Many Roofs (Phteah Dambaul Chraen

Figure 118 original Khmer tall house or "phteah khpasa" is a stilt house or pile house.

The "High House" (*Phteah Khpasa*) is ideal for the adventurous who like living on high stilts, giving ample storage room below. Originally from Srü, in the Kompong Thom province, situated some 150 km east of Siem Reap, it was built in 1985.

The original Khmer tall house or "phteah khpasa" is a stilt or pile house. The structure consists of evenly spaced wooden pilings that extend from the ground to the eaves or the roof ridge, historically called ridge-post framing. No provision is made at ground level for wall cladding or protection against wind and rain; by day, this area provides shade and serves as a living space for the inhabitants and their livestock. The upper floor, which is closed on all four sides, provides sleeping accommodation during the night; clothes, furniture and objects of value are stored here. A gable roof with an open truss completes the structure.

Figure 119 The "Fishmongers' House" (phteah anaknesaeat)

The "Fishmongers' House" (*Phteah Anaknesaeat*), built in 1958, was discovered in the village of Or in the Sang Voeuy commune. This village in the province of Siem Reap is situated some 60 kilometres from the town. Its owner, Mao Heng, is a retired fishmonger.

Depending on social status or financial possibilities, this basic structure may vary. According to the size of the building, further variations may be found by avoiding a great span width and economising on material; larger buildings no longer have a simple pitched roof. Such variations can often be seen in the partitioning of the upper floor: the division between the parents' room or the area set aside for the children may be extended to include a large veranda, thus providing a separation from the private parts of the house.

The "House with Horizontal Cladding" (*Phteah Kdab Phtek*) was Built in 1965; it was recovered from Yeang in the Russey commune. The village's name means bamboo. It is situated in the province of Siem Reap, approximately 50 kilometres from the town centre.

Figure 120 The House with Horizontal Cladding (Phteah Kdab Phtek)

Based on the framework and depending on the size of the building, 1 – 3 pitched roofs are placed alongside each other; the central roof will be noticeably smaller and narrower. A hip roof is another variation of the typical roof of a Khmer house; this construction requires a large amount of material and is complicated, so it is rarely seen. The shape of the roof defines the different house types. The Khmer House is an example of indigenous materials used in traditional vernacular architecture design.

Traditional houses in a settlement

Figure 121 Traditional house in a settlement

The house pictured here is situated in a settlement (Figure 117). Typical characteristics of such a settlement are: The private sphere of each household is restricted to the area occupied by their dwelling. Delimited by the stilt structure of the house, it is at ground level that daily life – work, eating and family interaction – takes place. Livestock is kept here in simple pens. These typical wooden buildings surround an open space (approximately 150 m^2), where a communal water source forms the centre of the settlement.

The images below show a few century-old houses the same family still owns. They have a wooden structure with bamboo walls. The roof is thatched with rice straw. The lifespan of such houses is quite long, although probably not more than 100 – 120 years. This house in the bottom right is 107 years old and reportedly still exists[82]. (See Figure 120)

Figure 122 Example Traditional Village Houses

Traditional Waterside Khmer House

These houses stand in water, temporarily or permanently, and are genuine pile buildings with no foundations. Wooden diagonal struts that strengthen the posts standing in the water ensure the stability of the building.

---

[82] https://en.wikipedia.org/wiki/Rural_Khmer_house

Figure 123 Traditional waterside Khmer house

There have been some developments using modern building materials, and the concept of building on stilts remains.

Figure 124 Further developments using modern building materials

Chapter 12

The basics of the Cham people

In the coastland of south-central Vietnam, most people are of Cham ethnicity. Their culture is diverse and rich because of the combination of indigenous cultural elements (plains culture, maritime culture, and mountain culture) and foreign cultural features... also "An Giang Province practices Islam and is also known as Cham, or Cham Muslims, around the region of Châu Đốc (An Giang Province) in the Southwest" (Pers. Comm. Pham Hong Thai, 2023).

The Cham villages are located on high land surrounded by farmland. Every clan or large household lives in houses in the shape of a rectangle or a square. Houses are built following a certain process.

A Cham village in Ninh Thuan Province

The Cham people do not grow tall trees, which they think are shelters for monsters. Thus, the Cham villages have few trees. A series of houses stand beside each other in the middle of parallel or perpendicular pathways. Doctor Le Duy Dai from the Vietnam Museum of Ethnology said: "Normally, the Cham houses face the East. In a Cham village, houses are nicely arranged like streets with alleyways. Village pathways are built in North-South or East-West directions. That's why the Cham houses or houses of a clan are built in a straight line".

In Cham's matriarchal family, every couple of sisters has their own house. So, the number of houses in a plot of land depends on how many married women that family has and on which class that family belongs to. Dr. Le Duy Dai said: "The aristocrats usually have seven houses sitting next to each other, while the poor have only five houses altogether. The extra two houses of the aristocrats are used to accommodate farming tools such as mandrels and shovels and household appliances such as grinders and rice-hulling mills".

Figure 125 The basics of the Cham people (Photo source: https://commons.wikimedia.org/wiki/Category:Cham_people#/media/File:In_front_of_Ch%C4%83m_house.jpg

The most important house among them is called *Thang do*, where family rituals take place like weddings and ancestor worshipping ceremonies.

Figure 128 A modified Cham House turned Bee Gallery.

"*Thang do* is the home of newly-wed couples. When the elder daughter gets married, she and her husband will live in *Thang do*, and her parents will move to another house called *Thang lâm*. When the second daughter gets married, the elder daughter's family will move to another house called *Mư dâu* and leave *Thang do* to her younger sister".

A Cham house in Ninh Thuan province, the Vietnam Museum of Ethnology[83]

Today, Cham's matriarchal system is changing into nucleus families. The number of houses is reduced, and the interior decoration is paid less attention. But the "soul" of each house remains:

Figure 127 An exhibit at the Vietnam Museum of Ethnology in Ninh Thuan province. Source: http://www.vme.org.vn/en

Figure 127 An outdoor exhibit at the museum, consisting of an Ede dwelling from the Central Highland region.

The Cham people choose the direction of the houses for particular reasons: "The direction of the houses is very important. For example, *Thang do* is usually in the East-North direction, considered the corner of the clan's reproduction. A traditional complex of 5 or 7 houses no longer exists. But many houses retain their structure. The house direction stays the same. For example, when people upgrade the *Thang tôn* house, its direction keeps facing the East. Similarly, the direction of Mư dâu remains unchanged although the house is renovated into a brick house with a flat roof or a multi-storey building".

One of Cham's most characteristic tangible cultural heritages and one of the most sensitive to change is their house. The Cham build their houses on the ground (with some having the floor raised on stubs) and arrange them in orderly rows. Their houses are surrounded by a garden yard with a wall or hedge.

---

[83] Source: https://en.wikipedia.org/wiki/Vietnam_Museum_of_Ethnology#/media/File:Dan_toc_hoc_19.jpg

The doors open to the southwest or between. The architectural style is similar to that of the Viet, with walls made of brick or a mixture of lime and shells and covered with tiles or thatch. Houses of more than one storey are rare. In certain localities, houses on stilts are found, but the floor is only 30 cm above the ground. The rooms of Cham houses are arranged according to a particular order: the sitting room, rooms for the parents, children, and married women, the kitchen and warehouse (including the granary), and the nuptial room for the youngest daughter. This arrangement reflects the break-up of the matrilineal extended family system among the Cham. The Cham living in Ninh Thuan and Binh Thuan believe they must perform certain religious rituals before building a new house, particularly praying to God and asking for his permission to cut down trees in the forest. When transported to the village, a ritual is also held to receive the trees. A ground-breaking ceremony called *phat moc* is also held. The precinct of the Cham traditional house is the residence site of a Cham family. It assembles several houses with different functions, which relate closely to each other.

Ref: http://en.vass.gov.vn/noidung/anpham/Lists/SachHangNam/View_Detail.aspx?ItemID=763

Islam is primarily the religion of the Cham people, an Austronesian minority ethnic group; however, roughly one-third of Muslims in Vietnam are of other ethnic groups. There is also a community that describes itself as of mixed ethnic origins (Cham, Khmer, Malay, Minang, Viet, Chinese and Arab) that practices Islam and is also known as Cham, or Cham Muslims, around the region of Châu Đốc in the Southwest.

Several Chinese accounts record Cham arriving on Hainan in 986, shortly after the Vietnamese captured the earlier Cham capital of Indrapura in 982, while other Cham refugees settled in Guangzhou. The **Utsuls** are a Chamic-speaking East Asian ethnic group living on the Hainan island. (https://en.wikipedia.org/wiki/Utsul)

Kampong Cham (Khmer: កំពង់ចាម; lit. 'Cham Port') is a province of Cambodia located on the central lowlands of the Mekong River. Kampong Cham is primarily lowlands. The main river is the Mekong River, which forms the province's eastern border, separating it from Tbong Khmum province.

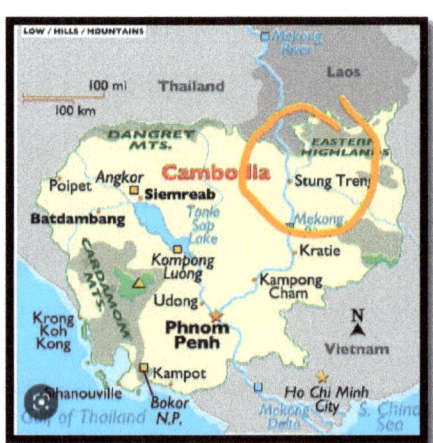

Figure 129 Location of the PEKABA visit.

A recent visit by the *Persatuan Kebajikan Amal Bangi* (PEKABA) or the "Bangi Charity Virtue Association" on one of their Charity drives provided some pics of the Kampong Cham traditional houses. (Courtesy of Datuk Abu Fitri)

Figure 131 Typical Kampong Cham Traditional Houses

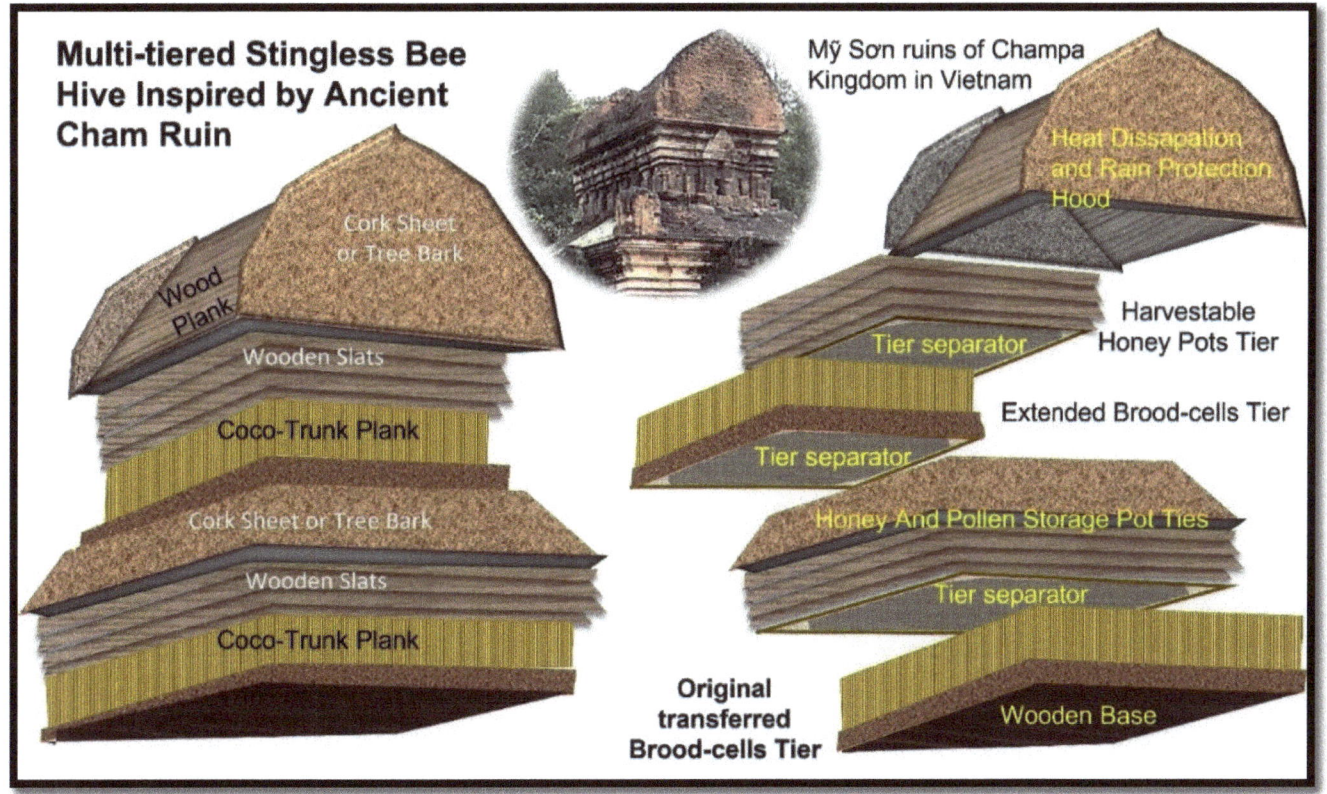

Figure 130 Bee hive design inspired by an ancient Cham ruin https://en.wikipedia.org/wiki/M%E1%BB%B9_S%C6%A1n

Mỹ Sơn (Vietnamese pronunciation: [mĭˀ səːn]) is a cluster of abandoned and partially ruined Hindu temples in central Vietnam, constructed between the 4th and the 14th century by the Kings of Champa, an Indianized kingdom of the Cham people. The *Mỹ Sơn* temple complex is regarded as one of Southeast Asia's foremost Hindu temple complexes and is the foremost heritage site in Vietnam.

Hani Homes and Mushroom Houses

Hani houses are either built on the ground or raised about 1½ meters off the ground on low stilts. A typical house has a thatch roof, walls made with logs or bamboo and a storage area for produce below the living quarters. Many houses are in a fenced compound containing the main house and a granary and smaller houses used by extended family members. Each house has a masculine and feminine side and a space for pigs. Each house has an ancestor altar.

Figure 133 In places like Honghe, Yuanyang and Luchun, houses have mud walls (bricks, made from mud or clay) and thatched roofs supported by wooden pillars placed on stone foundations.

A mushroom house, as the name implies, is a mushroom-shaped house made of earth or mud walls, bamboo and wood framing and a straw roof with four slopes. The ground floor of the house is for keeping livestock and furniture. People live on the first floor, which normally has three rooms and a fire pit with a constantly burning fire. The second floor is covered with fireproof earth and serves as a warehouse. Not only pretty when seen from afar, the mushroom house is also noted as being warm in the cold winter but cool in the hot summer.

Figure 132 A Hani house in Vietnam Source: https://en.wikipedia.org/wiki/Hani_people#/media/File:Nh%C3%A0_ng%C6%B0%E1%BB%9Di_H%C3%A0_Nh%C3%AC_(m%E1%BA%B7t_b%C3%AAn).jpg

[Source: Liu Jun, Museum of Nationalities, Central University for Nationalities, Science of China]

The Hani or Ho people (Hani: Haqniq; Chinese: 哈尼族; pinyin: Hānízú; Vietnamese: Người Hà Nhì / 𠊛何貳) are a Lolo-speaking ethnic group in

Figure 134 20111102-wiki com Akha Hut.JPG

Southern China and Northern Laos and Vietnam. A typical house of this kind (Figure 130) is found in the largest Hani village – Mali Village of Yuanyang County, Honghe City. According to the legend, the Hanis lived in caves in ancient times. Later, when they migrated to "Reluo", they found a vast field of mushrooms all over the mountain and the plains. The mushrooms could resist wind and rain, and ants and worms built nests under them. The Hanis imitated the mushrooms and built mushroom houses (Figure 131).

The base of the wall is built of stones or bricks, half a meter high both above and under the ground. Upon the base is the wall, made of earth, pressed hard between boards. The architectural design of the two or three-floor mushroom house is very special and charming. The roof is covered with multi-layers of couch grass, shaped into four slopes. The inside usually comprises the principal room, the front veranda and the penthouse. The front veranda is connected to the front wall of the principal room, and the penthouse is connected to one or two sides of the principal room. The front veranda and the penthouse's roof are solid soil platforms where people rest and enjoy the cool or dry-reaped crops. The second floor of the principal room is completely walled with clay, and the roof is covered with grass three or four meters above the floor. The space inside is called the "Fire Banking Tower", which is divided with boards to store crops, melons and beans. Also, there is a place for young people of the right age to spend the night.

The lowest floor is used to keep livestock and to pile up farm tools. The middle floor is separated into three rooms, with a square fire pit burning through the year. When guests come, the host sits by the pit and offers puffs from a bong-like bamboo water pipe, cups of hot "glutinous rice tea", and bowls of

Figure 135 Green roofed Hani 'Mushroom House'

sweet "sealed wine". Mushroom houses are durable, warm in the winter and cool in the summer. They look charming, grouped together in a village perched on a mountain slope amid green staircases of rice terraces.

Some thatched roofs have flowering creepers, enhancing aesthetics and providing thermal comfort.

List of Records of the Stingless Bee Type localities in Vietnam (Rasmussen 2008):

Homotrigona anamitica (Friese, 1909)

Trigona anamitica Friese 1909("1908"): 358-359, fig 15-4: Lectotype (ZMHB, worker): here designated, "Süd-Annam / Xom-Gom / February / H. Fruhstorfer", "Trigona / anamitica / 1904 Friese det. / Fr."; paralectotype (ZMHB (1)) (comparative note, taxonomy); **Type locality: VIETNAM** "Xom-Gom, SüdAnnam, February 1901 (Fruhstorfer leg.)" (3 workers);

"Trigona (Lepidotrigona) ventralis" nigribasis forma nov. Sakagami 1975: 71: Nomen nudum. The name is unavailable for taxonomic purposes, as it must be deemed infrasubspecific (ICZN, art. 15.2 and 45.6.3.) (variation); VIETNAM.

Lisotrigona carpenteri Engel, 2000

Lisotrigona carpenteri Engel 2000: 232-235, 236, figs. 1-3: Holotype (IEBR); 3 paratypes (AMNH, IEBR) (illustration, key to species, taxonomy); **Type locality: VIETNAM** "VIETNAM: Nghê An Province: Khe Bo, 19 03N 104 43E, 25-28 April 1998, 123 m, James M. Carpenter" (worker); VIETNAM, Ha Thing, SE Huong Son;

Tetragonula gressitti (Sakagami, 1978)

Trigona (Tetragonula) gressitti Sakagami 1978: 166-194, 206, 214-216*, 226, 227, 228, 234, 235, 236, 237, 238, 239: Holotype (BPBM; not located, nor in SEHU); paratypes (SEHU) (taxonomy); Type locality: VIETNAM "Fyan-b, Viet Nam" (49 workers, holotype, paratypes); "Fyan-a" (1 worker, one male); "Dalat" (5 workers); Balao (1 male);

Records of the Type locality in Cambodia:

Tetragonilla collina (Smith, 1857)

Trigona cambodiensis Cockerell 1926a: 224: Holotype (USNM 29470, worker): examined (nest, taxonomy); **Type locality: CAMBODIA** "Angkov (=Angkor) Wat, Cambodia, Jan. 11, 1926 (H.M. Smith)" (4 workers); Cockerell 1929b: 140 (distribution, nest); CAMBODIA, Angkor; Kum Puang Creek; Nan;

Tetrigona apicalis (Smith, 1857)

Trigona hemileuca Cockerell 1929b: 140: Holotype (BMNH 17b.1098) (comparative note, taxonomy); **Type locality**: THAILAND "Siam: Nan, Dec. 31 and Jan. 31 (Cockerell)" (worker); **CAMBODIA**, Ban Maa Hia; Mekami river;

Tetrigona peninsularis (Cockerell, 1927)

Trigona apicalis peninsularis subsp.n. Cockerell 1927: 541: Holotype (BMNH 17b.1099) (taxonomy, variation); **Type locality: MALAYSIA** "Perak, F.M.S., Batang Padang, Jor Camp, 1800 ft., June 4, 1923 (Pendlebury)"; **CAMBODIA**, Patalung; MALAYSIA, Kuala Lumpur, Gombak valley;

Meliponiculture in Vietnam

by Viet Thang, interviewed by AHJ (2016)

Vietnam is a tropical country with flora that produces abundant nectar. It has great prospects for research and breeding of stingless bees. However, there is little information about various aspects of stingless bees and almost no rearing of stingless bees or meliponiculture.

According to available information (Chinh 2005), Meliponines in Vietnam exist in three genera: *Lisotrigona, Tetragonula* and *Lepidotrigona*. It contains information about the architecture of the hive and colony characteristics of *Lisotrigona carpenteri*, *Lepidotrigona ventralis* (eggs in layered composition) and *Tetragonula laeviceps* (eggs in clustered cells). In addition, new information on the production of sexuals (queens and males) and what extranidal and intranidal factors are associated with this process in the species *L. ventralis*, a layered comb builder. There is also interesting information about how the seasonal production between drone and gyne plays an important role in revealing the mechanism of colony multiplication of this species (Chinh 2004)[84].

Meliponiculture

Interview with Viet Thang (Admin Member of fb group Stinglee Bee – Vietnam) https://www.facebook.com/groups/812385295534445/ STUDENT at National Economics University
Well, I have 30 colonies of stingless bees. There are two species in Vietnam, and I keep one among them. I harvest manually. I practice two harvesting methods: With the craft knife and vacuum pump.

The farm location is here –
https://www.google.com/maps/place/12%C2%B029'11.8%22N+109%C2%B006'55.6%22E/@12.4866222,109.1132523,17z/data=!3m1!4b1!4m5!3m4!1s0x0:0x0!8m2!3d12.486617!4d109.115441
I'm in Thuan Thua, 468km northeast of Ho Chi Minh City.

The rate of failure is 20% to 30%. I failed to divide the colony (colony splitting). But after 45 to 60 days of care, I continue to get the rest back successfully. It is also the most difficult job of taking care of them.

---

[84] The article in mention is "Research on stingless bees in Vietnam (Apidae, Meliponini) [2004] Chinh, T.X., Bee Research and Development Centre, 68 Nguyen Hong, Lang Ha, Dong Da, Ha Noi (Vietnam). "
Source: University Library, University of the Philippines at Los Baños
Homepage: http://www.uplb.edu.ph

Figure 136 a) b) & c) Meliponine hive station; d)storage pots in the bamboo hive; e) brood cells in the bamboo hive; f) & g) preparing to transfer the colony; i) transfer into a box.

Very few farmers are in my area, raising four people, including two small farmers, with only three colonies and me. I cannot get the close-up image because it needs a good camera. I got the first stingless bee honeycomb in one bamboo population at tourist resorts. Stingless bees live in bamboo tubes there. I do not make hives of bamboo, which is a natural thing. Possible to harvest 360ml to 460ml of honey collected in the honeycomb bamboo. I started stingless bee breeding farms on 11/2015

Figure 137 a) Log hive; b) nest entrance; c) brood cells clustered; d) High sugar content in honey; e) Provision of high-quality honey retains aroma, colour and taste after refrigeration; e) pot-pollen harvest

Model making for bee housing

Figure 138 Tuấn Vũ of National Association of Beekeeping [HỘI NUÔI ONG DÚ TOÀN QUỐC] https://www.facebook.com/groups/2134975603386858/user/100002286322548

References

Chinh, T. (2004). "Research on stingless bees in Vietnam (Apidae, Meliponini). *Bee Research and Development Centre, 68 Nguyen Hong, Lang Ha, Dong Da, Ha Noi (Vietnam)*.

Chinh, T. X., & Sommeijer, M. J. (2005). Production of sexuals in the stingless bee *Trigona (Lepidotrigona) ventralis flavibasis* Cockerell (Apidae, Meliponini) in northern Vietnam. *Apidologie 36*, 493-503. doi:10.1051/apido:2005035

Chinh, T. X., Sommeijer, M. J., Boot, W. J., & Michener, D. C. (2004). Nest architecture and colony characteristics of three stingless bees in North Vietnam with the first description of the nest of *Lisotrigona carpenteri* Engel (Hymenoptera: Apidae, Meliponini). *Journal of the Kansas Entomological Society 80(2):* 130-135.

Rathor, V. S., Rasmussen, C., & Saini, M. S. (2013). New record of the stingless bee *Tetragonula gressitti* from India (Hymenoptera: Apidae: Meliponini). *Journal of Melittology No. 7*, 1–5 6.

Chapter 13
Perspectives on Meliponiculture in SE Asia
By Stephen Petersen (akbeeman2000@yahoo.com)

Introduction

Stephen Petersen has spent the last 12 years travelling and consulting on bee development projects from Africa to SE Asia. He has been involved in development projects in Laos, Cambodia, Malaysia, Myanmar, Thailand, Rwanda and Uganda, plus observed in many other countries. He spends the summers in Alaska with his own bee business (25 years of experience) and, escaping the cold and dark, travels to warmer climes in winter to pursue his love of bees and their role in development projects.

What is meliponiculture?

Meliponiculture is the management of stingless bees to benefit from their pollination efforts or harvest

Figure 139 Stingless bee box hive

a small amount of high-value honey.

Here in SE Asia, there are more than 50 species in the Tribe Meliponini inhabiting a variety of habitats.

What's in a Name?

Local names

Malay = kelulut; Khmer (Cambodia) = maroam; Thai south = ong; Lao & Issan = kee suit; India = dammar bee

The 'Entrance Tube', composed of wax and plant resins, is one of the signs of a stingless bee colony. They will live in various containers and can be easily managed for pollination or producing small amounts of high-value honey.

The colony is similar to honey bees (*Apis*); brood, pollen and honey are stored separately but in small "pots" instead of hexagonal cells.

Worker Development in *Tetragonula iridipennis* (stingless bees)*

Egg - *4.7 days*

Larva - *18.5 days*

Pupa - *21.3 days*

Total - *44.5 days*

(About 2X times longer than honey bees)

* Data courtesy of Kerala Agricultural University (India) for *Tetragonula iridipennis*.

The Queen is the most important bee in the colony

- ❖ She mates with only one drone on her nuptial flight.
- ❖ She is much larger than the other colony members and, therefore, easy to spot.
- ❖ Her wings partially cover the abdomen.

Whole nests can be removed from forest locations, brought to a central location, or relocated into rustic log hives.

Purpose-made containers can simplify management and harvesting

Large-diameter bamboo serves as a nesting site.

A purpose-made box for efficiently managing *Trigona*[85] bees – the interior partition in the shape of a "T" allows the separation of brood cells and honey pots.

Harvesting honey from stingless bees

Left – using a bamboo knife to separate the honey pots from the wall of the log hive. Right – a kitchen spoon is used to scrape out honey pots. In each case, a container is used to catch the drips!

Separating the honey from the wax

The honey pots (above) can be broken open (be careful to eliminate any pollen pots), squeezed, passed through a fine mesh filter, and then bottled.

Figure 140 a) b) c) assorted nest entrances d) Internal nest structure e) Tetragonula sp. f) Tetragonula queen g) established box hive h) brood cells, honey & pollen pots.

Figure 141 a) log hives newly placed b) log hives hung up c) Bamboo hive d) purpose-made wooden box e) manual harvesting f) box hive diagram g) Honey from box hive.

Beekeeping in Laos and Cambodia

Review by Abu Hassan Jalil

I could not communicate directly with any farmers or beekeepers in Laos or Cambodia, and I can only describe what I have comprehended from the following article. Development of beekeeping in LAOS: Various strategic choices. by Bounpheng Sengngam, Professor of Phytopathology to the Faculty - Agriculture of Nabong, National University of Laos, Email : bounpheng_sengngam@yahoo.fr

and Jérôme Vandame, Agronomist of the Comité de Coopération avec le Laos (CCL), Technical assistant of the Projet d'Appui à la Faculté d'Agriculture (PAFA), Email : jerome_vandame@yahoo.fr

The article is mainly an outlook on Apiculture in Laos and also looks at Meliponiculture as an alternative.

Thus, various species of stingless bees, whose surface of the extension includes Laos, may have a future in the agricultural systems of Laos.

The colonies of stingless bees can colonize various spaces:

• Basement, ant-hill: in particular, the case of *Tetragonilla collina*, which seeks an environment with low thermal amplitude.

• Hollow tree trunk. In this case, the species *T. collina* and *Homotrigona fimbriata* choose these spaces because of their low thermal amplitude.

• Tree trunks of small diameter and branches of trees. It is, in particular, the case of the species *T. laeviceps* that adapts itself to temperature variations.

The installation in a hive of the first two species is delicate due to the ineptitude of these bees to adapt to the variations of temperatures of these closed spaces. On the other hand, it is very easy to install a colony of *Tetragonula laeviceps* in a hive.

Production and pollination

This species has, besides, a strong faculty for pollination, which completes the interest given by the production of small quantities of honey and pollen, both very much wanted because of gustative quality for the first one and their energizing quality for the second. In addition, the products of this rustic beekeeping could be developed in the niche markets of Laos and neighbouring countries because of their specificity and great intrinsic quality.

In the zones of fruit production, beekeeping (meliponiculture) with *Tetragonula laeviceps* is a credible alternative. In addition to the interest in improving pollination and thus the quality and quantity of the fruits obtained, these kinds of beekeeping must allow the production of small quantities of honey and pollen appreciated in Southeast Asia for their medicinal properties.

Another article of interest is "Diversity of stingless bees (*Hymenoptera, Apidae, Meliponini*) from Cambodia and Laos" by Seunghwan Lee et al.

"In the recent investigation, fourteen species of stingless bees are recognized in the dry season from Cambodia and Laos, which are congeners to the species from an adjoining country, Thailand. Three species of stingless bees are recognized for the first time from Cambodia: *Pariotrigona pendleburyi* (Schwarz, 1939), *Tetragonula sirindhornae* (Michener and Boongird, 2004) and *Tetrigona melanoleuca* (Cockerell, 1929). Likewise, two species of stingless bees were reported for the first time from Laos: *Lisotrigona furva* Engel, 2000 and *Tetragonula fuscobalteata* (Cameron, 1908). Images of morphology and nesting behaviour, a checklist of stingless bees from Cambodia and Laos and a short discussion on biology are provided."

No. Taxon (Meliponini)	Cambodia,	Thai	Laos	Viet
1. *Geniotrigona thoracica* (Smith 1857)	KP	*	*	*
2a *Homotrigona fimbriata* (Smith, 1857) [*aliceae* (Cockerell, 1929)]	widesprd	*	-	-
2b *Homotrigona fi mbriata* (Smith, 1857) [*anamitica* (Friese, 1908)]	MK	-	*	*
3 *Lepidotrigona terminata* (Smith, 1878)	widesprd	*	*	*
4 *Lepidotrigona ventralis* (Smith, 1857) [=*fl avibasis* (Cockerell)]	MK	*	*	*
5 *Lisotrigona cacciae* (Nurse, 1907)	MK, SI	*	*	*
6 *Lisotrigona carpenteri* Engel, 2000 (Fig. 3b)	MK	-	*	
7 *Lisotrigona furva* Engel, 2000	MK	*	*	-
8 *Tetragonilla collina* (Smith, 1857)	widesprd	*	*	*
9 *Tetragonula fuscobalteata* (Cameron, 1908)	widesprd	*	*	*
10 *Tetragonula geissleri* (Cockerell, 1918)	KK, MK	*	*	-
11 *Tetragonula* sp. [cf. *laeviceps* (Smith, 1857)]	PP?, SI?, ST?	*	*	*
12 *Tetragonula pagdeni* (Schwarz, 1939)	KC, KK, PP7	*	*	*
13 *Tetrigona apicalis* (Smith, 1857)	MK, ST	*	*	*

However, it is uncertain how the bees were identified. Some notes were made regarding this in "The Cambodian Journal of Natural History". In his article in the said Journal, John Ascher mentions many errors in identifying "Diversity of Stingless Bees from Laos and Cambodia" by Lee et al., 2016 accepted version.

The John Ascher et al. article from pages 23 -39 Cambodian Journal of Natural History 2016 touches on two colour forms of *Homotrigona fimbriata* [the flavinic *aliceae* (Cockerell) in western Cambodia and the melanic *anamitica* (Friese) in eastern Cambodia] and cites *flavibasis* in synonymy as the name available for Cambodian *Lepidotrigiona ventralis* (Smith) *sensu lato* if considered a distinct species.

Stingless Bees In Phnom Penh

Interview of Zohosy Haroun by Abu Hassan Jalil

A stingless bee enthusiast from Phnom Penh recently visited me. Zohody Haroun, known as Soty (back in his village), was advised by the Malaysian Embassy in Phnom Penh, Kemboja, to visit my meliponiculture setup. Soty, a Cambodian national, worked at the Malaysian Embassy and was on vacation.

He treated me to some of his videos and photographs from village areas in Phnom Penh (https://www.facebook.com/abu.h.jalil.1/videos/10211124667783422/;

https://www.facebook.com/abu.h.jalil.1/videos/10211124666503390/;

https://www.facebook.com/abu.h.jalil.1/videos/10211124725984877/), one was at a termite mound and a few on sed posts. They appeared to be a tiny species and most likely of the genus *Tetragonula*.

Figure 142 Typical Tetragonula sp. Nest in a shed post.

This species even makes nests in trees in front of the village houses. Those at forest fringes have trees occupied by meliponines of bigger sizes, and honey hunters have destroyed most. These hunters will climb trees up to 15 feet(5M), armed with axes or machetes, will chop at the trunk where they locate the stingless bee nest and cut out a big portion of the tree trunk to get at the nests in the cavities. They are very reckless and are not concerned about the remnant nest or the condition of the trees. Much destruction is left after they have had their fill.

Figure 143 Small stingless bees make nests in village trees

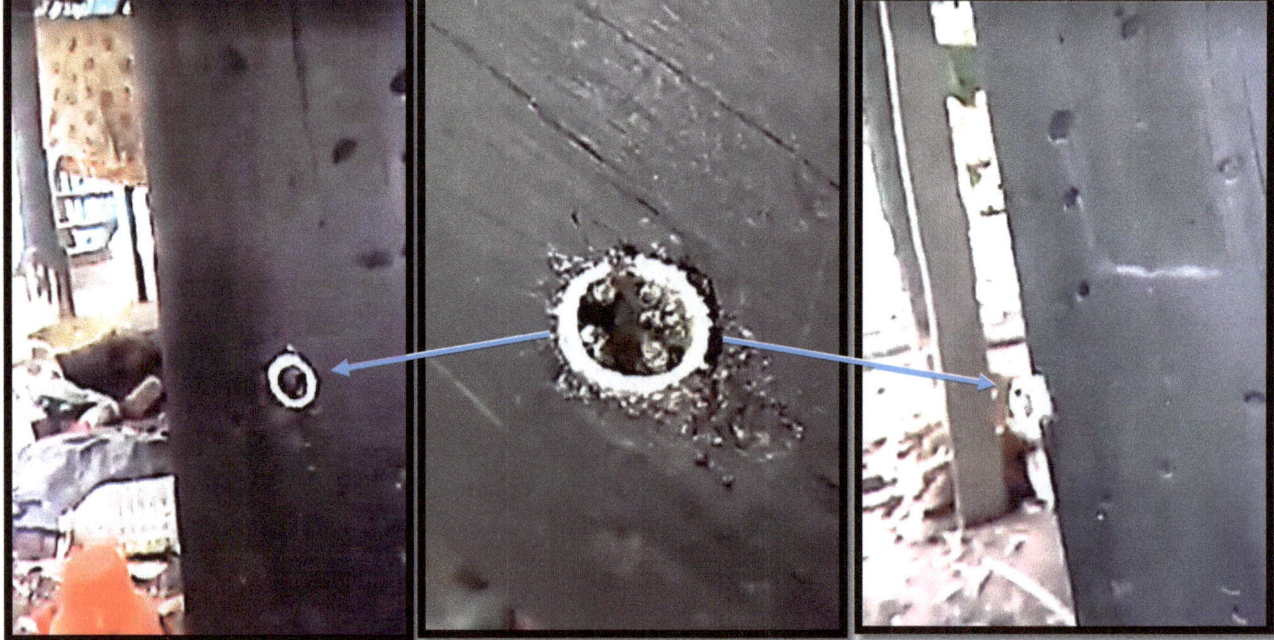

Figure 144 Nest entrances found in wooden House posts

References

Ascher, J. S., Phallin, H., Sokha, K., Kang, L., Sokchan, L., ShaoXiong, C., . . . Sophany, P. (2016). A report on the bees (Hymenoptera: Apoidea: Anthophila) of Cambodia. *Cambodian Journal of Natural History*, 23-39.

Couvillon, M. J., Wenseleers, T., Imperatiz, V. L., & Noguiera-Neto, P. (2007). A comparative study in stingless bees (Meliponini) demonstrates that nest entrance size predicts traffic and defensivity. *J . EVOL. BIOL. 21 (2008)*, 194–201. doi:10.1111/j.1420-9101.2007.01457.x

Jump, D. (2016). *Stingless Bees in NW Cambodia*. Retrieved from Bees Unlimited: http://www.beesunlimited.com/trigona-photo-album

Jump, D. (2016). Cambodia's passionate "bee-doctor". (Inga, Interviewer) Retrieved from http://www.visit-angkor.org/blog/2015/10/02/interview-with-dani-the-passionate-bee-doctor-in-cambodia/

Lee, S., Duwalb, R. K., & Lee, W. (December, 2016). Diversity of stingless bees (Hymenoptera, Apidae, Meliponini) from Cambodia and Laos. *Journal of Asia-Pacific Entomology, 19*(4), 947–961.

Sengngam, B., & Vandame, J. (2003). Development of beekeeping in LAOS: Various strategic choices. *Bees for Development*.

Sakagami, S. (1978). Stingless bees (excl. *Tetragonula*) from continental Asia and Sri Lanka (Hymenoptera: Apidae). *J. Faculty of Science, Hokkaido University Series VI. Zoology, 21*, 165- 247.

Sakagami, S. F. (1978). *Tetragonula* Stingless Bees of Continental Asia and Sri Lanka (Hymenoptera, Apidae). *Journal of The Faculty of Science Hokkaido University Series VI.Zoology, 21(2):* 165-247.

Sakagami, S. F., Yamane, S., & Hambali, G. G. (1983). Nests of Some Southeast Asian Stingless Bees. *Bull. Fac. Educ., Ibaraki Univ. (Nat. Sci.)*, 1-21.

Petersen, S., & Bryant, D. M. (2010). The Role of Melissopalynology in the Honey Market. Hanoi: Texas A&M University.

Rahman, A., Das , P. K., Rajkumari, P., Saikia, J., & Sharmah, D. (2015). Stingless Bees (Hymenoptera:Apidae: Meliponini): Diversity and Distribution in India. *International Journal of Science and Research, 4 (1)*, 77-81.

Vijayakumar, K., & Jeyaraj, R. (2014, Oct 26). Taxonomic notes on stingless bee *Trigona (Tetragonula) iridipennis* Smith (Hymenoptera: Apidae) from India. *Journal of Threatened Taxa, 6*(11), 6480-6484.

Chapter 14

Ancestor veneration among some Asian cultures

Especially in the North of Chiang Rai province, there is a native tribe of the Akha people. Akha religion—*zahv*—is often described as a mixture of animism and ancestor worship that emphasises the Akha connection with the land and their place in the natural world and cycles[1]. As we trace them westwards to Laos, mountain villages occupy close quarters in Houaphanh Province. The roofs in Laos have a taller pitch and do not have the 'scissors' finial of the Thai Akha at the gable ridge ends. Interestingly, they have the old Malay Kampung house's similar gable & hip roof.

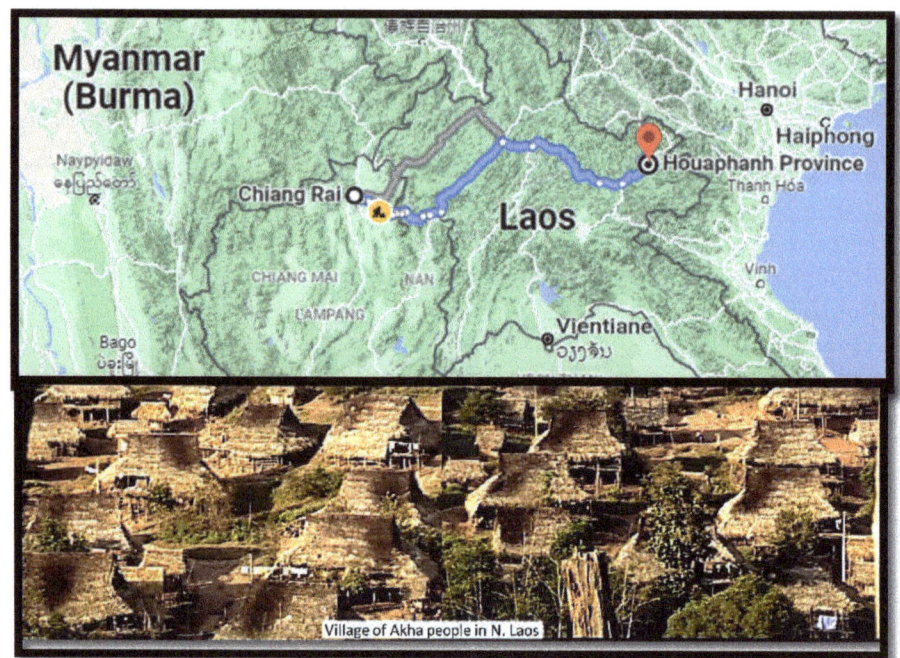

Figure 1 Location of Akha Tribal villages

The Akha are an ethnic group who live in small villages at higher elevations in the mountains of Thailand, Myanmar, Laos and Yunnan Province in China. They went from China into Southeast Asia during the early 20th century. A civil war in Burma and Laos resulted in an increased flow of Akha immigrants, and there are now 80,000 people living in Thailand's northern provinces of Chiang Rai and Chiang Mai. (Source: https://en.wikipedia.org/wiki/Akha_people)

Further south, the Buddhist traditions prevail in all the southern regions of Indo-China, from Burma, Siam, Laos, and Cambodia to Vietnam. Anyway, back to Laos's Akha culture, some influences could be from the Yunnan province up north in China. The Daen Lao Range; Burmese: Loi La is the Shan Hills Mountain range in eastern Burma and northern Thailand. Most of the range is in Shan State, which borders China (Yunnan) to the north, Laos to the east, and Thailand (Chiang Rai, Chiang Mai

---

[1] Ref. video: (https://www.youtube.com/watch?v=o384Ezk3HrY Laos Wonderland (full documentary) - The Secrets of Nature)

and Mae Hong Son Provinces) to the south. We suspect there is some influence on ancestral veneration from South China.

Figure 2 Akha Tribe in Chiang Rai province near the Thailand & Laos Border.

Figure 3 Inspired by the Karen Hill tribe traditional architecture.

Other tribes that may practise ancestor veneration in the region are the Karen hill tribe of Chiang Rai, Thailand and the Tai Dam or Black Tai, an ethnic group (sub-group of Karen) found in parts of Laos and Vietnam. Figure 143 shows examples of a traditional Tai Dam house, Ban Na Pa Nat Tai Dam Cultural Village, Loei Province in NE Thai (close to Vientiane, Laos) and a split-level house of the Karen tribe at a hill slope in Chiang Rai province.

The Karen is one of the largest hill tribes in Southeast Asia, and there are said to be approximately 400,000 Karen Hill tribe people living in the hills of Northern Thailand. The Karen have been known to use elephants to help clear the land for farming and have since established a reputation as elephant trainers or mahouts. There are several subgroups of Karen tribes, the most common being the "Dam" (Black). The language used by various subgroups is different, but since all originate from Sino-Tibetan ethnicity, they can communicate. The Karen is often confused with the Red Karen (long neck), famous for the neck rings worn by women to stretch the neck. The Karen people have traditionally practised ancestor worship, and the village is overseen by a Chief or spiritual leader, who has great power and control of the local community.

Figure 4 Inspired by a Tai Dam (Black Tai) tribal house in Laos.

The Tai Dam (Lao: ໄຕດຳ, Thai: ไทดำ) are an ethnic minority predominantly from China, northwest Vietnam, Laos and Thailand. They are part of the Tai people and ethnically similar to the Thai from Thailand, the Lao from Laos and the Shan from Shan State, Myanmar. Tai Dam means "Black Tai". This name comes from the black clothing the group wears, especially females. In Vietnam, they are called Thái Đen and are included in the group of the Thái people, together with the Thái Đỏ ("Red Tai"), Thái Trắng ("White Tai"), Phu Thai, Tày Thanh and Thái Hàng Tổng.

Influences from China

Ancestor veneration practices prevail in South China, where lineage bonds are stronger in the patrilineal hierarchy – meaning they tend to worship male ancestors. So, looking at South China architecture shows upsweep in the roof eaves as we saw in Vietnam.

Figure 5 A study of the architecture of the ancestor veneration practices prevailing in South China.

Xiazetang Village in Jinxiang Town, Cangnan County, Zhejiang, China. In the family temple of the Tang 汤 family and the village cultural and recreational centre, the upsweep in the roof eaves shows how the belief system of ancestors in the heavens above brings the eaves up instead of the usual downward flow for rainwater drain off. The ridge dragon ornament symbolises the heavens. This influence is quite obvious in important buildings[2] in Vietnam.

Still, on the track of ancestor worship, the Hokkien Chinese (Southern Min language originating from the Minnan region in Southeastern Mainland China) celebrated their 9th day of Chinese New Year last night (the author was working on this chapter during the Chinese New Year 2022). This celebration is rampant throughout Malaysia and Singapore.

Figure 6 South Chinese architectural influence in North Vietnam. https://www.alotrip.com/about-vietnam-culture/brief-history-vietnam-architecture

---

[2] Ref: https://www.alotrip.com/about-vietnam-culture/brief-history-vietnam-architecture

Figure 7 Hokkien Chinese (Southern Min language originating from the Minnan region in Southeastern Mainland China) celebrated their 9th day of Chinese New Year last night. This celebration is rampant throughout Malaysia and Singapore

The photos on the left are the burning of house replicas (middle photo) to send to their ancestors in the heavens. In Singapore, they also burn sugarcane[3].

On the right is an ancestral worship ceremony led by Taoist priests at the pyramidal-shaped Great Temple of Zhang Hui (张挥公大殿 Zhāng Huī gōng dàdiàn), the central ancestral shrine dedicated to the progenitor of the Zhang lineage[4].

Figure 8 Examples of a pyramidal roof on Box hives

Coincidentally, this pyramidal roof shape is used as a bee box hive roof. However, it has nothing to do with the Chinese ancestor worship culture. These were done in reflection of the Javanese Joglo. The Japanese aesthetics, however, do not appreciate the pyramid shape. Although it is stable in equilibrium, they are still boring. The Japanese aesthetic principles call for rhythm and flow. The ancient code of bearing a weapon from left to right is implanted in the flow and rhythm of their nature.

---

[3] Further reading: https://mothership.sg/2022/02/why-do-people-burn-sugarcane-on-9th-day-of-lunar-new-year/
[4] https://en.wikipedia.org/wiki/Ancestor_veneration_in_China#:~:text=
Chinese%20ancestor%20worship%20or%20Chinese,into%20lineage%20societies%20in%20ancestral

The Japanese artistic ideal is the isosceles mountain shape (esp. ratio 3:5:7), steeper on the left and gradual on the right.

Other forms of Ancestor worship are found on an island between Taiwan and the Philippines. The Ivatans[5] of Batanes Island of the Philippines practised ancestor veneration. Today, however, most Ivatans are Catholics, like the rest of the country, although some have not converted and practised ancestral worship to their anitos[6].

Before the Spaniards arrived in the Philippines, Ivatans built their houses from cogon grass. These homes were small, well-situated, and designed to protect against strong winds. The Spaniards introduced large-scale lime production to the Ivatan to construct their now-famous stone houses. The basic cogon grass is still preserved as roofs of their houses, thickly constructed to withstand strong winds. Meter-thick limestone walls are designed to protect against the harsh Batanes environment, which is known as a terminal passage of typhoons in the Philippines.

Figure 9 A Sinadumparan Ivatan house is one of Batanes Islands' oldest structures. The house is made of limestone and coral, and its roofing is cogon grass.

More of Filipino vernacular architecture in Part 1 of Volume 3 – Philippines Architecture, where we touch on pre-colonial housing and architectural developments in the Spanish era.

---

[5] https://en.wikipedia.org/wiki/Ivatan_people#/media/File: Oldest_House_in_Ivatan.jpg
[6] Anito, also spelled anitu, refers to ancestor spirits, nature spirits, and deities in the indigenous Philippine folk religions from the precolonial age to the present, although the term itself may have other meanings and associations depending on the Filipino ethnic group.

VOLUME 2, PART 3
~ Climate Adaptation & Fusion Architecture ~

Introduction to Part 3

This part briefly encompasses some climate change adaptation and fusion architecture and touches on ancestor veneration among some Asian cultures, including the Philippines and Himalayan states. The first few chapters in this third part deal with thermal imaging of Indochinese homes to show the coolest areas in typically traditional houses. This may help determine the cooler spots to place beehives.

The subsequent chapters deal with combining traditional or folk architecture with colonial influence. Examples are the Sino-Portuguese and the Dutch and other European influences, including Moorish designs. After this is a treatment to Ancestor veneration accents that have made for unique vernacular designs that may serve well in tourist attractions and indigenous skills in each culture's construction. These skills are found peculiarly interesting among skull collectors of the indigenous peoples of Borneo, Papua, Northeast India and the Philippines.

Finally, it was necessary to provide a separate treatment for Miniature landscapes to accommodate the miniature model making of ASEAN houses used in vernacular meliponiculture in the last chapter.

Chapter 15

Changing Climate

Thermal comfort in a tropical and sub-tropical climate

Cambodia lies in a tropical zone, 10 to 13 degrees north of the equator. The monsoon cycle dominates the climate, with dry and wet seasons. **November to February** '*cool and dry*'; **March to May** '*hot and dry*'; **June to August** '*hot and wet*'; **September to early November** '*cool and wet*'.

Temperature and humidity recordings

Temperature and humidity recordings show variations in temperature of the inner and outer surfaces are no greater than + / - 1 °C. Thermal imaging confirms this, which leads to the conclusion that there is no noticeable difference in temperature outside or inside the house. Contrary to expectations, constant air circulation, which would have a cooling effect on surface temperatures, does not occur. However, the gap between the roof and the walls allows air to flow in and out, preventing heat accumulation below the roof.

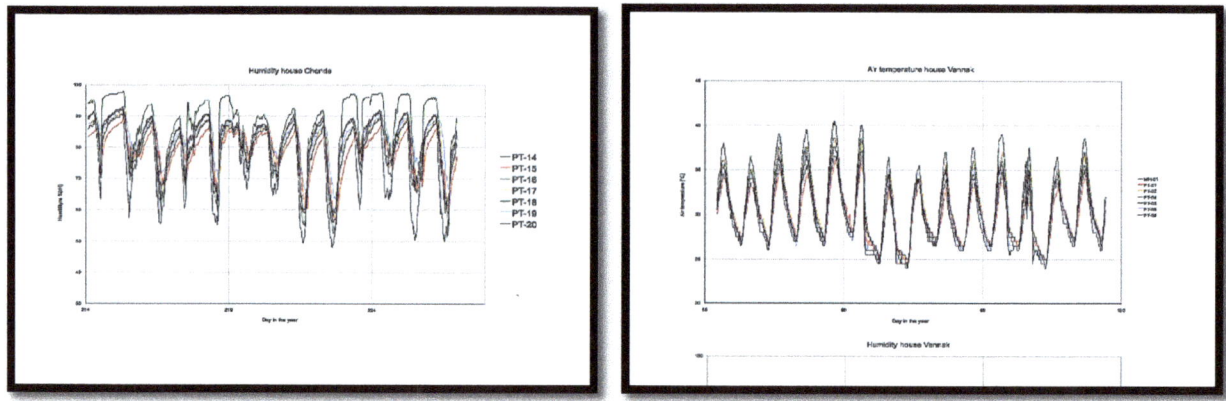

Figure 154 Temperature and humidity recordings

Thermal imaging

Infrared thermal imaging reveals that the lowest temperatures are found at ground level beneath the house. It can be said that the comfortable indoor climate is brought about by combining the 'umbrella effect of the roof and natural ventilation. The roof shields the house against direct heat radiation, whilst the continuous opening

Figure 155 Thermal imaging

between the roof and the top of the walls allows a constant flow of air in and out, which prevents an accumulation of warm air beneath the roof.

Regional Heat Wave[92]

Vernacular Architecture is a Preventative measure to keep our bees safe from heat waves.

Myanmar: The temperature at Myinmu, a river town west of Mandalay, rose to its highest at 47.2 °C (116.96 °F) on May 14 as a moderate El Niño occurred in 2010

Thailand: Mae Hong Son saw the hottest temperature on record in Thailand, topping out at 44.6 °C (112.3 °F) on April 28, 2016. Uttaradit held the previous record at 44.5 °C (112.1 °F) on April 27, 1960.

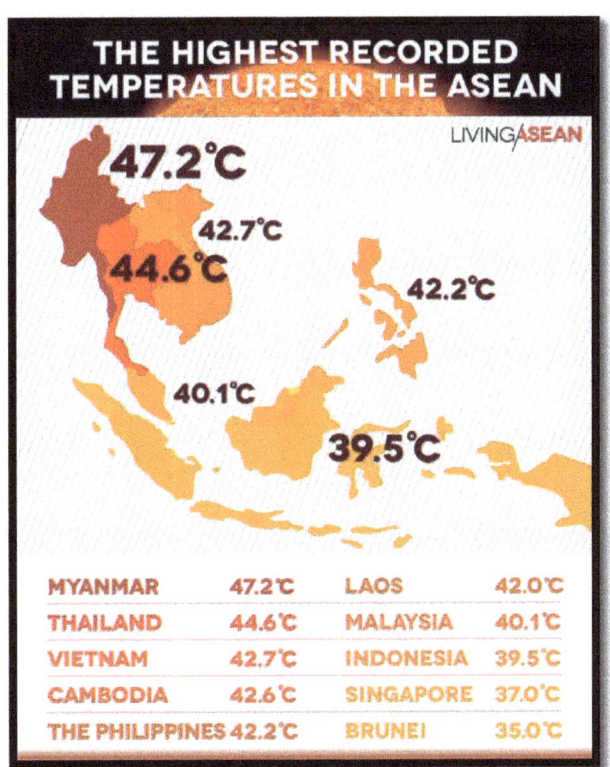

Figure 156 Highest Recorded Temperatures in the ASEAN

---

[92] Source: https://livingasean.com/special-scoop/highest-recorded-temperatures-asean/?fbclid=IwAR1bAGOksTKX2px9FbTB2MqQwNUW6iUHTFiEnK6k0C5Kql7EZAJ5k8EmB2U

Vietnam: The temperature at Con Cuong rose to a record 42.7 °C (108.9 °F) on May 30, 2015. The city of Con Cuong is in Nghe An Province in the North Central region of Vietnam.

Cambodia: The new record high for Cambodia was set on April 15, 2016, in Preah Vihea when the mercury soared to 42.6 °C (108.7 °F).

Conclusion

Despite maintaining a comfortable, draught-free environment, indoor temperatures of 36 °C, up to 80% relative humidity, and no electric air-conditioning are possible thanks to the umbrella effect of the roof.

Recent Global Heat Waves

Web Ref:	1, 2, 3 & 4	5 & 6	5 & 7	8 & 9	10 & 11
Temp.	41	45 C	50 C	110 F	109 F
Region	The UK.	India	Pakistan	China	Texas, USA

1. https://www.youtube.com/watch?v=Z3WIssUqe6s
   A heat wave across the world News 19 WLTX - 20 Jul 2022
   High temperatures are making their way across Europe.

2. https://www.youtube.com/watch?v=nADhWieuFLc
   Climate Expert On Extreme Heat: 'We're Not Going To Be Able To Find Solutions
   MSNBC- 20 Jul 2022
   Climate scientist Dr. Sweta Chakraborty, writer-at-large at The Bulwark Tim Miller, NBC News Correspondent Emilie Ikeda, and NBC Foreign Correspondent Meagan Fitzgerald join Chris Jansing to discuss the climate change-induced, record-breaking heatwave engulfing Europe. "We need to mitigate against worst-case warming scenarios because, at some point, we're not going to be able to find solutions to this runaway heat and to the many impacts of climate change that are ripple effects from this temperature warming," says Dr. Chakraborty.

3. https://youtu.be/S91HEBv_Gf0
   Europe's heatwave spreads north as wildfires rage in the south | DW News - 28 Jul 2022
   Europe is going through another summer of record-breaking temperatures. The north is bracing for a week of extreme heat, while the south is already experiencing temperatures climbing to new highs. Hundreds have died from the effects of the extreme heat. Strong winds fuel destructive wildfires across several nations, and several firefighters have been killed.

4. https://youtu.be/75GnrwDgmko

   The UK records highest-ever temperature as fires spark across London Global National |Global News: July 19, 2022

   A record-breaking heat wave has turned Britain's capital into a tinderbox, with fires erupting across London, prompting the city to declare a major incident.

5. https://youtu.be/SEnBtdmhSsk

   How much is climate change to blame for heatwaves in South Asia? | Inside Story Al Jazeera English - 1 May 2022

   Scientists have long warned that climate change will lead to more intense weather. Many points to record-breaking temperatures now in India and Pakistan as proof. Although heatwaves are common in the region, they don't usually start until May or June. The intense heat has increased the electricity demand, leading to power outages. So, what can be done to reduce the impact of heat waves?

6. https://www.youtube.com/watch?v=AuuPTqZBd18&t=2345s

   India's Extreme Summer: Will 45°C Heatwaves Leave Us Out Of Wheat? | Insight: CNA Insider - 26 Jul 2022

   India experienced an unprecedented heatwave this year. As temperatures regularly soared past 45°C in northern India, the wheat-producing region's crops wilted in the fields. Indian farmers saw their yields plummet, leading to debt and ruin. At the same time, prices of wheat and related products skyrocketed, and ordinary consumers bore the brunt of rising inflation. Food security is at risk.

   India eventually moved to ban wheat exports, but there are concerns that the heat wave is not an anomaly but an increasingly regular occurrence caused by climate change. Can Indian farmers weather this crisis, and is this a sign of things to come?

7. https://youtu.be/iVhcaQMVa2M

   Pakistan Burns At 50°C: Can It Turn The Heat Down On Climate Change? | Insight CNA Insider 2 Aug 2022

   A brutal heatwave has hit Pakistan. The scorching heat wave has transformed several cities into hell on Earth, with some of the country's daily temperatures hitting 50 degrees Celsius. It has destroyed crops and ice sheets much faster than previously anticipated. Officials have warned of acute water shortage, and the country is starting a major food crisis. Pakistan is now ranked the 6th most vulnerable country on the planet, and its Ministry of Climate Change estimates that climate change could cost the country anywhere from $7 billion to $10 billion a year in disaster response alone, never mind the massive losses in economic activities caused by the intense heat. Can something be done to reverse the tide of climate change before it is too late? And will the people be able to fight the biggest threat to their existence?

8. https://youtu.be/oJxLRKHtQgY

    Temperatures soar across China & Japan | Heatwave prompts power crunch warning | WION (World is One News) - 30 Jun 2022

    Japan has been reeling under the worst heat wave that the country has seen since 1875. Hundreds of people were treated for heat stroke as the temperatures in Tokyo broke nearly 150-year records for June. Meanwhile, China has also issued a yellow alert for high temperatures and heat waves in various parts of the country.

9. https://www.youtube.com/watch?v=zAln0kO5J9s

    Gravitas: Why factories in China are shutting down. WION (World is One News) - 20 Aug 2022
    Scorching temperatures across the country, and its longest river is running dry.

10. https://youtu.be/OYC2Mra6pPs

    What's The Deal With These Heat Waves? | Neil deGrasse Tyson Explains... StarTalk
    3 Aug 2022

    What is a heat wave? In this explainer, Neil deGrasse Tyson and comic co-host Chuck Nice explore the massive heat waves sweeping the world with the Director of NASA Goddard Institute for Space Studies, Gavin Schmidt.

    Learn what qualifies something as a heat wave. We break down the extreme global heat and the pressure it's been putting on current infrastructure. We enter the jet stream and wave patterns in the Earth's atmosphere. Why exactly are we seeing more extreme heat? Why do we pay attention to extreme heat in some regions but not others? Will some places become uninhabitable? Finally, learn about the last ice age and our climate's changes. What's our heat limit?

    A question posed is, "Why is the UK overly concerned? The Queen is there!"
    This caption sets us to think of our queen bees.

11. https://www.youtube.com/watch?v=Y6u14FGwPwA

    Texas Sees 3 Weeks Of Triple-Digit Temps With No Relief In Sight |TODAY - 5 Aug 2022
    Texas is bracing for its relentless heat wave to continue as temperatures are expected to break 100 degrees again on Friday, marking three straight weeks of triple-digit highs. NBC's Sam Brock reports for TODAY, and Al Roker has the latest forecast nationwide.

Chapter 16

Out of the box into the sphere

By Abu Hassan Jalil

Saturday night, I am reminiscing about the 70's Disco sphere with glittering psychedelics. The millennials are clueless about this. I call this chapter "Out of the box into the sphere," or is it out of the circle but still in the square? The readers can decide. I am looking at a beekeeper in North Malaysia

Figure 157 Windmills as his Meliponary theme by KT Chan of Alor Janggus, Kedah, Peninsular Malaysia.

who took the windmill as his Meliponary theme.

KT Chan of Alor Janggus, Kedah, Peninsular Malaysia, created his colourful Meliponary to attract tourists. But is this vernacular architecture? Well, to the Hollander, it is. I guess the intricate carvings are too cumbersome. The vanes can rotate but only manually. Including an interior mechanism to provide air circulation within the "attic" would be great.

The next structure is truly out of the box. It is more in line with yoga practitioners. A custodian, Freddie Lozada of the Gawahon Eco Park in Victorias, Negros Occidental, showed me this structure he intends to build out of G.I. pipes using pipe benders.

Figure 158 Conceptually for yogic meditation, but looks like a good protective cover for box hives during typhoons

Such a marvel idea for many beehives to be protected from strong wet winds. I searched for more images online and came up with some modifications.

The following design may appeal because you are more comfortable working with straight beams.

The hives can be kept safe and comfortable with the back to the winds. However, it has not been tested in a Typhoon. This design is out of the Vernacular Box but still in the Bee comfort Sphere. If you like it, I have attached the details herewith.

Figure 159 An example covering all weather

I prepared this next one for a friend in Gawahon Eco Park in N. Occ. Negros. The Gawahon Eco Park is already a Tourist attraction with its native birds and other flora and fauna. Adding this Vernacular architecture to their bee housing will elevate their attraction to a higher level.

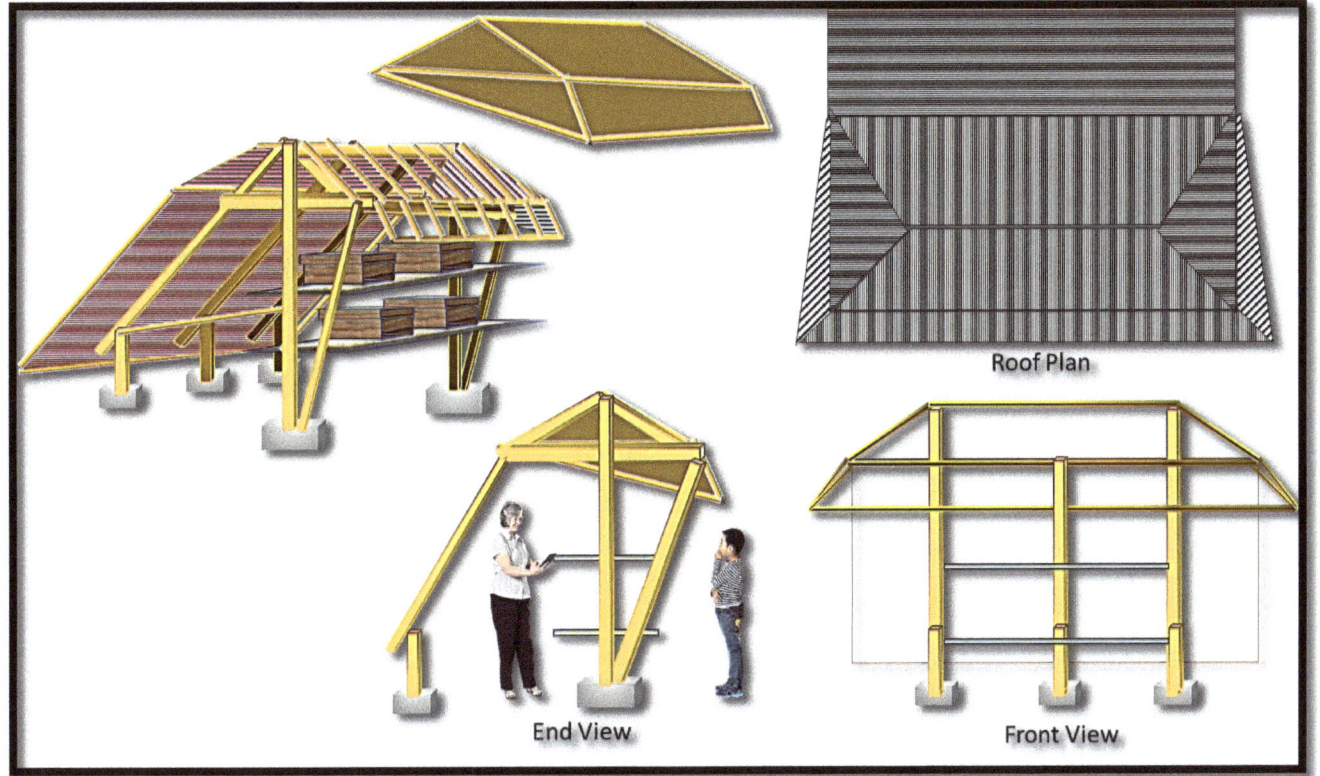

Figure 161 Inspired by the bus shelter concept to shelter beehives

For measurements, you can scale to the human figure height. The second shelf is at a kid's eye level. The posts can be anchored with mass concrete. The concrete anchors can be attached to ground beams for added stability. Small moats can be incorporated with used engine oils to deter crawling insects.

Figure 160 Anchor by footing and ground beam for maximum stability.

These bases should stabilise the flimsy structure during earth tremors and strong gales.

Chapter 17

Fusion of Traditional architecture
By Abu Hassan Jalil

I have come full circle from where I started in Langkawi Island on Insect Tourism. Down the Straits of Malacca, through the Java Sea to the Sulu Sea. From Sulu to the South China Sea through the Gulf of Thailand. Moving across the Isthmus to the Andaman Sea and back to Langkawi. Now, we see Langkawi as a melting pot of cultures. The Thai Influence from the north and northeast, Arab and Indian Influence from the northwest, Acehnese from the west and finally southwest, and the Malays in the east and southeast.

I have always wondered why Mr. Nasiruddin chose this roofing style for his bee box hives in his Meliponary behind the Langkawi Cultural Centre, Langkawi Island. Malaysia.

Figure 162 Left: Mr. Nasiruddin chose this roofing style for his bee box hives in his Meliponary behind the Langkawi Cultural Centre: the ornately carved bee box; right: a proposed bee house to blend into the Langkawi landscape, the concrete stairway anchors the whole structure at high tides.

Aceh roots and Thai blended into the Traditional Malay House have become typical Langkawi traditional homes. Moorish-style arches indicate some Arabic influence.

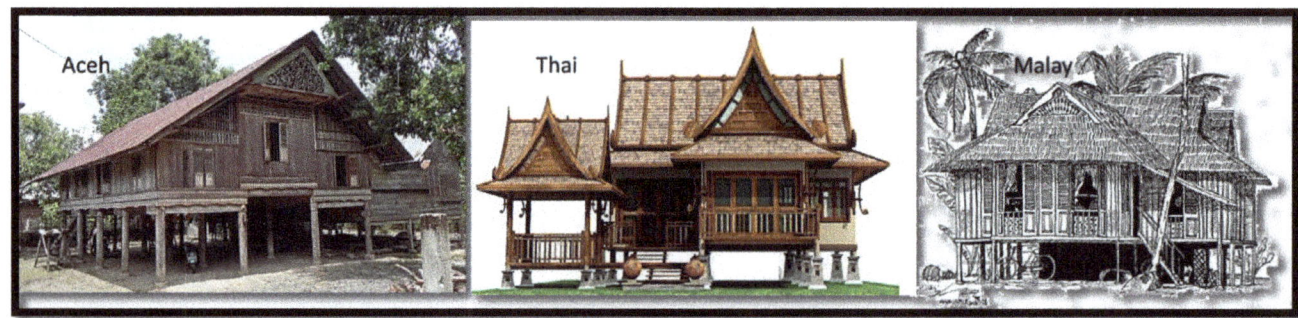

Figure 163 With Aceh roots and Thai blended into the Traditional Malay House, these have grown into the typical Langkawi traditional homes.

The far right image of Figure 147 is a bee house that blends the influencing cultures. The concrete stairway anchors the whole structure at high tides for those affected when built in the coastal areas.

I remember more than 40 years ago my first visit to Langkawi (my honeymoon, actually). I was invited to lunch at a coastal fishing village in Kuah town. The host merely went down to his 'backyard' the sea and netted some fish as the tide rose. They scaled and gutted the fish directly into the frying pan. What a blast it was, having freshly fried caught fish for lunch.

His house front faces the ground-level road, and the side towards the back, he had his boat docked facing the sea. After lunch, we got to his boat and accompanied him to work. A wonderful and memorable experience. Anyway, the houses inland are quite different.

Figure 164 A typical fishing village row of houses.

The Langkawi Island inland houses are on stilts, too, but the stairway entry is usually incorporated within the whole building roof structure system. This feature is evident in the traditional Aceh architecture.

Figure 165 The stairway is usually receded into the building under the main roof structure.

Sino-Portuguese and European Fusion
By Abu Hassan Jalil

The Portuguese influence was overwhelming in the Straits settlements, namely Penang, Malacca and Singapore. Besides the Moors, the Dutch and the British also had a hand in shaping Malaysian Architecture.

Figure 166 This Governor's Mansion in Phuket City has the Malay Gable and Hips style roof with a partial Hexagonal central foyer-like section.

This Governor's Mansion in Phuket City (Figure 151) has a Malay Gable and Hip-style roof with a partial Hexagonal central foyer-like section. The Istana Jahar Palace 19th century wooden building is now Kota Bharu Malaysia's Museum (Figure 152). We can see that it is essentially oriental with upswept ridges at the eave's corners. It uses the Malay Gable and Hip roofs with some European colonial pressure at the central porch.

Figure 167 The Istana Jahar Palace 19th century wooden building is now Kota Bharu Malaysia's Museum.

A familiar style of this Istana Jahar Palace is seen in the typical Sarawak Malay House (see Figure 38).

Istana Ulu (Early Palace of Perak) in Kuala Kangsar, Perak Darul Ridzuan, Malaysia, incorporates many ornamental elements of mansion designs in the Sino-Portuguese style. It has prompted me to incorporate the 'castle tower roof' in a box hive rack. I include the wind vane and a dormer to house a ventilator fan if needed.

Figure 168 *Istana Ulu (Early Palace of Perak) in Kuala Kangsar, Perak Darul Ridzuan, Malaysia, incorporates many ornamental elements of mansion designs in the Sino-Portuguese style.*

This next one combines the curved pitch of a Dutch barn roof or bonnet roof (like the Lumbung of Lombok Island) with the 2nd pitch skirting incline of 30° like the Malaccan 2nd pitch roof. Looks like a Dutch Bonnet. Incidentally, Malacca and Lombok were colonised by the Dutch at one time or another.

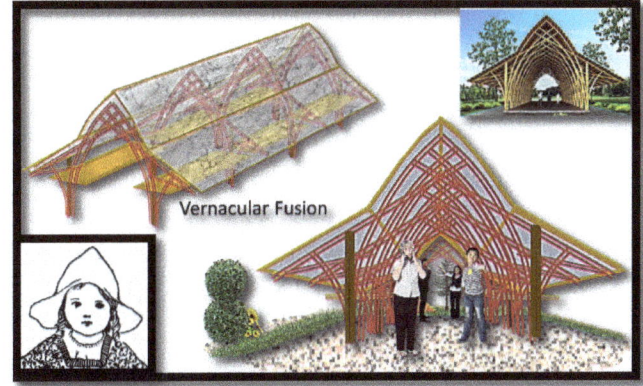

Figure 169 *Vernacular fusion of the Bonnet roof with a 2nd pitch skirting. Bottom left inset: A Dutch doll bonnet.*

Chapter 18

Diversity of Indo-Malayan Vernacular Architecture

In the age of climate change and variations, we strive to achieve the ultimate stability in thermal comfort and equilibrium in our foundations to outlast tremors, floods and gales. We look back to our ancestors' brilliance revealed in past structures and how to tackle extreme weather and climates. We look at collective comparisons of variations as changes may be new in an area while they may be known already in another area. We learn from other regions where one extreme was once common and how the indigenous peoples overcame that extreme. Any particular extreme may be new, and we need to apply what we can learn from others with experience.

Chinese influence encompasses the north of Central Asia down to Laos and Vietnam. North of Myanmar and Thailand, the indigenous tribes remain with their own culture (pagan culture and animism) to the south. The Khmer influence adhered to Hinduism and Buddhism until it carried to Sri Vijaya and Majapahit, like Borobudur relics in Java. Malay architecture flourished in the Isthmus of Kra in the Malay Peninsula and directly flows to Borneo and the Indonesian archipelago. Then, it also clashed with the influence of the Polynesian islands in NTB and NTT but saw the Dutch influence in Lombok and Maluku and the Portuguese influence in Flores and the east (See Chapter 10 p.126).

However, the Sultan of Bima was not perturbed, and there was still an element of Malayness in Sumbawa and Bima. The Sultanate of Bima (كسلطانن بيما) was a Muslim state in the eastern part of Sumbawa in Indonesia, at the site of the present-day regency of Bima. It was a regionally important polity that formed the eastern limit of Islam in this part of Indonesia and developed an elite culture inspired by Makassarese and Malay models.

In Mindanao, the Moro Malay influence collided with the Spanish influence. In Palawan and Sulu, the influence of Borneo Malays and the sultan of Sulu still exists.

A touch of the 'Head Hunter' vernacular accent

By Abu Hassan Jalil

In reminiscing about my past during my bachelor days in the '70s, I used to have a friend from the Iban tribe (a denomination of the Dayaks, the native people of Sarawak). She was working and living in Kuala Lumpur, and with the stigma of her ancestry, she did not make many friends. The friendship

Figure 170 Inspired by the Orang Ulu tribe of Sarawak.

remained platonic because of the different religious faiths. They were headhunters in the past. The Christian missionaries and Raja Brooke of Sarawak (Sir James Brooke, the first White Rajah of Sarawak) eradicated the headhunting practice in the 19th century.

Orang Ulu ("people of the interior" in Malay) is an ethnic designation politically coined to group roughly 27 small but ethnically diverse tribal groups in north-eastern Sarawak, Malaysia. The Orang Ulu tribal groups are diverse; they typically live in longhouses elaborately decorated with murals and woodcarvings. They are also well known for their intricate beadwork, detailed tattoos, rattan weaving, and other tribal crafts. Most of the Orang Ulu tribes are Christians now, but old traditional religions are still practised in some areas.

When I ventured to Borneo a few years ago, I knew I would be meeting many different Dayak tribes of the four corners of Borneo. I want to talk about East Kalimantan, the Kenyah Dayak and The Kutai Dayak tribes (not to be associated with the *Rumah Kutai* in Perak State). This passage connects the headhunting tribes of Indo-Malaya with 'Nusantara', the proposed name for the new Indonesian capital. I visited Samarinda Beekeepers in East Kalimantan. You wouldn't know their native ethnic origin when you meet them, although some are proud of their heritage displayed in their artwork. The beekeeping is done in villages on forest fringes, and the vegetation provided was much desired.

I explained to the East Kalimantan beekeepers that although they are near the forests, one cannot expect much productivity if the bees are not foraging the forests. The bees have particular preferences and affinity for foraging. They have since beefed up their respective bee farms with Beescape.

Figure 171 Ornately carved bee box hive; Right: AHJ and Agus Suhartanto with his assistants.

Agus Suhartanto and friends in Tenggarong, North of Samarinda, carved these bee box hives. That's where I met 'Loloq', Agus' pet owl. Loloq keeps his bees safe from swiftlets, swallows, frogs, lizards, rats, small snakes, and occasionally large spiders, although he will not fly into their web.

Figure 172 AHJ with Loloq, the pet owl.

While in East Kalimantan, I also visited the Samarinda Ethnic Culture Museum by night, and with that, I proposed a box hive rack for them.

Agus carved an assortment of hilts for the Mandau (a Dayak headhunters' machete, see Figure 62) on the left. Some hilts of the past were made from deer horns. The middle photo is the totem poles at the Museum and a "Belawing" next to it, which may serve as a clan identity 'mascot' and the hornbill.

Figure 173 Left: Mandau hilts carved by Agus; Mid left: Totem poles at the museum; Mid right: a Dayak Belawing (clan identity pike); Right: Bee rack with traditional Kutai Dayak motifs.

On the trail of Indo-Malayan Head-hunter Tribes

The **Ibans** were the greatly-feared headhunters of Borneo. Headhunting among the tribe is believed to have begun when the land inhabited by the Ibans grew into overpopulation, and confrontation was an inevitable requirement for survival. The head hunters in Borneo were active until approximately one century ago. Various tribes, including Sarawak's Iban, Sabah's Murut and Kadazan-Dusun, brought the early British colonialists fear.

The **Igorot** people are Austronesians. They were known in earlier days for their wars and practice of headhunting. The Spaniards forcibly partially subdued them during the colonial occupation of the Philippines. That process was completed during the period of U.S. hegemony[93].

July 3 2018, and head hunters of the **Ouroun** people of Papua New Guinea. Goaribari is an island in southern Papua New Guinea. Indeed, the Indonesian Papua is proudly a land of headhunters – notorious tribal warriors, and there were many alarming cases of strangers being attacked, killed and eaten by the indigenous people. West Papua is still a field of the possibly last surviving tribes engaging in cannibalism.

Catholic missionaries, many with degrees in anthropology, were partially successful in persuading some of the Asmat to stop cannibalism and headhunting. Headhunting was evidenced by the discovery of thousands of skulls in village houses and the longhouse in the early 20th century. When headhunting

---

[93] Etymology - From the post-classical Latin word hegemonia (1513 or earlier) from the Greek word ἡγεμονία hēgemonía, meaning "authority, rule, political supremacy", related to the word ἡγεμών hēgemōn "leader".

culture meets ancestor veneration, they don't strictly collect skulls but were once cannibalistic (as late as the 60s), so they have leftover skulls after ritual feasting of the brain and the lower mandibles are made into necklaces for the women. Ancestor skulls, however, are extremely sacred and are kept well-decorated. Strongly believe spirits of their ancestors dwell within their totem carvings. The Asmat has three basic colours in its décor: white from lime, black from charcoal and red from ochre clay-like terra cotta.

There are the people of the fascinating **Konyak** tribe, better known as the head hunters. They are the largest of the 17 officially recognised tribes in Nagaland and are famed for their age-old tradition of headhunting. Head-hunting was practised among the **Mizo**, the **Garo**, and the **Naga** tribes of India, Bangladesh, and Myanmar until the 19th century. The Konyaks of India's northeast practised headhunting into the 1960s. Some of those warriors are still alive today.

For further reading:

1. https://anthrosource.onlinelibrary.wiley.com/doi/pdf/10.1525/aa.1959.61.6.02a00080
2. https://www.newworldencyclopedia.org/entry/Headhunting
3. https://kashgar.com.au/blogs/tribal-culture/the-practice-of-headhunting
4. https://theculturetrip.com/asia/malaysia/articles/the-history-of-borneos-headhunters/
5. https://www.flyingdusun.com/004_Features/041_headhunting.htm
6. https://www.outlookindia.com/outlooktraveller/explore/story/47456/malaysia-the-tribals-of-sabah

In 1998, I got a Landscaping contract with Labuan Municipality, an Island off the coast of Sabah. I established a plant nursery with a few **Murut** and **Kadazan-Dusun** descent employees. The last we heard of headhunters was as late as WWII when the practice was reactivated among some tribes. Oh no, no! I was not fearful but always cautious, and neither was I obsessed with headhunters. I was and still am intrigued by the skull collections.

I still wonder how they clean and preserve their heads when returning to their longhouses. The stench must have been awful for months on end. I asked my Kadazan boys about it. They said they traditionally scalp the head and remove the brains and fluids through the nostrils and the occipital region (nape of the neck). After that, they would smoke the head until it was dry and the facial skin shrunk tight and then use camphor to camouflage the aroma... The old skulls will not have any skin left and are dark in colour, not all bone-white skulls as we would expect.

There are cultural villages for tourists in Sabah (https://www.borneotravel.com/blog/sabah-borneo-cultural-villages). One can get all the information there. As it turns out, my boys pulled a fast one on me. The boys told me their fathers practised headhunting, more like their grandfathers' fathers. Hah! Grandfather's tale!

Dr. Roubik may have something to say about this; one of his favourite topics is necrophagous bees[94]. If I were them (knowing what I know now), I would use necrophagous insects (stingless bees included), which they probably did, but without records, we do not know about them. We know of *Trigona hypogea* and its group *Trigona crassipes* and *T. necrophaga*... vulture bees or carrion bees.

We do not have *Trigona hypogea* or members of its group in Indo-Malaya, but I will tell you how I came to learn about carrion foraging bees. I was travelling back from the Mt. Ophir (Gunung Ledang) expedition with Mohd Noor Isa of MGVI and Dr. Hans Bänziger, and we were chatting in the car about the best food for our bees. Dr. Bänziger replied one word "cadaver!". I was stunned and asked, "What cadaver?" Mohd Noor replied, "A corpse," They all laughed at me and said the cadaver would have all the minerals and proteins essential for the bees. So, what's the best food? Note to self: If you want to learn, let your mentors laugh at your ignorance occasionally.

Looking back, I remember seeing bees in the wall crevices of an old tomb in the Jewish cemetery and Tombstone crevices in a Chinese cemetery on Penang Island, where I grew up. Later, when I was hive hunting about a decade ago, I found five nests in trees near a Malay cemetery in Kelantan. Seems like the bees don't care about your race or religious faith. If you are dead, they will make nests nearby. I have seen *Heterotrigona itama* foraging rotting meat once, but I have never encountered stingless bees burrowing in any graveyard. They must detect some scent favouring cavities to nest in graveyard vicinities.

On Timor Island, Indonesia, the **None** tribe of West Timor used to practice head-hunting until the 1940s. In the eastern region of Timor Leste, there are remnants of an Ancestor veneration culture where the inhabitants keep all their ancestors' belongings in a sacred house called Uma Lulik.

---

[94] (Systematics and bionomics of the apoid obligate necrophages: the *Trigona hypogea* group (Hymenoptera: Apidae; Meliponinae) João M. F. Camargo, David W. Roubik Biological Journal of the Linnean Society, Volume 44, Issue 1, September 1991, Pages 13–39, https://doi.org/10.1111/j.1095-8312.1991.tb00604.x

Chapter 19

Skull Collector Tribes

I append below some illustrations of known head-hunting skull-collector tribes: **Igorot** of the Philippines, **Asmat** of Papua, Indonesia, **Kadazan** of Sabah, Malaysia and **Sema Naga** of Nagaland, India and the **Aztecs** with their skull-collecting antics.

In this chapter, I have included illustrations of the **Orang Ulu** Longhouse (a subgroup of Iban Dayak) of Sarawak, the **Korowai** treehouse of PNG and the **Murut** longhouse of Sabah. They have built houses on stilts to avoid wild animals and also from conflicting enemies and some river pirates. So, now that I am designing bee housing, I am inspired to work on headhunters' domicile architecture as well.

The gruesome practice of taking human heads is particularly associated with the **Igorot** peoples of the Cordillera of Luzon. These all engage in it or have done so until recently. But today, the most persistent and dreaded head-hunters are neither Igorot nor inhabitants of the Cordillera; they are a wild, forest-dwelling people in the broken and almost impenetrable mountain region formed by the junction of the Sierra Madre range with the Caraballo Sur. They have been called different names by the peoples contiguous to them on the north, west and south, "Italon," "Ibilao," "Ilongot", or "Ilūngūt". The last designation would, for some reason, be preferred, but "Ibilao," or as it is quite commonly pronounced locally through northern Nueva Ecija, "Abilao," has perhaps the widest use.

Ilongot is known locally as the Bugkalot. The Ilongots were known for their distinct tribal accessories and headhunting tradition. For tourists, some displays and an Ilongot Tower replica can be found in Nagtipunan.

Ilongot Headhunting Tradition Bladed weapons of the Ilongot were made from traditional agricultural and hunting tools. Those used for warfare or headhunting were usually made for the purpose. Adangsel[95] introduced us to the *iteng*, a short blade resembling a scythe used for *kaingin* and cutting grass. For head hunting, warriors used the *tagyaden*, a fierce-looking curved *bolo*. It is used with a

Figure 175 The Ilongot or Ibilao house of Luzon and the Ilongot Tower

precise backhand swing to chop the head of the enemy. The blade handle and sheath are decorated with beadwork, tassels, and metallic embellishments.

The headhunting tradition among the early Ilongot was a form of expression of grief. During grief, the tribe would go on a killing spree. Severing the heads of a member of an enemy tribe and an outsider meant acquiring an amet, the spirit of the beheaded that was necessary for emotional balance and cultural regeneration. This ritual was also practised when the human head was offered a wedding dowry.

Gaddang Tribe of Solano

Figure 174 Gaddang Tribe of Solano

---

[95] https://traveleronfoot.wordpress.com/tag/ilongot/

The Gaddang lived High in the mountains in forested areas, and it was difficult for people to get to where they lived. They lived in houses built up off the ground on poles. Some of them also lived in tree houses. This helped to keep them dry when it rained. They used bamboo to build their walls and thatched roofs. They lived near a stream for reliable water supply and near their fields, usually on the slopes of a valley. More literature on Filipino Pre-colonial Indigenous architecture is in Part 4.

Skull collecting in Aztec culture

Figure 176 An Aztec Golden Pavilion and a skull rack (from https://en.wikipedia.org/wiki/Aztecs#Aztec_culture)

Diverting from the Indo-Malayan Eco Zone, we examine the Aztec skull-collecting antics. The depiction of a tzompantli ("skull rack"), the right half of the image, is associated with the depiction of an Aztec temple dedicated to the deity Huitzilopochtli. From the 1587 Aztec manuscript, the Codex Tovar.

Description from World Digital Library: "This illustration, from the second section, shows (at left) a temple or pyramid surmounted by the images of two gods flanked by native Mexicans. On the temple is an image of Huitzilopochtli on the right, and an image of Tlaloc holding a turquoise serpent is on the left. The temple is surrounded by a wall of serpents swallowing one another's heads. At the right is a tzompantli (Aztec skull rack). Huitzilopochtli, meaning "Blue hummingbird on the left," was the Aztec god of the sun and war. The xiuhcoatl (turquoise or fire serpent) was his mystical weapon. Tlaloc, the god of rain and agriculture, was of pre-Aztec, or Toltec, origin. A coatepantli (wall made of sculpted serpents) often surrounds Aztec temples. The tzompantli would hold the skulls of sacrificial victims. The great temple at Tenochtitlan was surmounted by two sanctuaries—the one on the left dedicated to Tlaloc, the one on the right to Huitzilopochtli."

Most recent headhunting news

The last instance heard of a headhunting event was a clash of ethnic cultures in Central Kalimantan in 2001. The Sampit War was one of the most terrible wars between peoples in this archipelago.

This incident occurred when the hostility between two peoples in Central Kalimantan, namely the Dayak as natives and the Madurese as immigrants, peaked. The riots and killings between these two groups, which have been going on for a long time, finally led to this tragedy on February 18, 2001, in Sampit City, Central Kalimantan, Republic of Indonesia.

It is said that in this war, the Dayaks have revived the hereditary tradition of their ancestors, namely "*ngyau*", meaning "hunting for heads", a tradition of the Dayak people who are most surrounded by

Figure 177 Image source https://en.wikipedia.org/wiki/Longhouse and https://en.wikipedia.org/wiki/Dayak_people

all nations in this region. As a result, about 100 bodies of Madurese were found lying headless. Ref: https://youtu.be/RTI-0X4hvJ0

Pak Ancah, Desa Penggilingan Padi, Sampit, Central Kalimantan

I was visiting a friend in Sampit in 2018. Memories of the past war have since long passed. My friend, Indra, brought me to the village. It was still drizzling when we arrived at Penggilingan Padi village, about a half-hour south of Sampit town. Suppose we can see the nests with the umbrella and the powerful flashlight that Indra brought him. The size of the topping box is a bit too big, in my opinion (approx. 40 x 40 x 16cm). They are mostly atop *H. itama* about 10 log nests

Figure 178 AHJ with the family of Pak Ancah. Penggilingan Padi, Sampit

around the house. The plants around the house are a few fruit trees with a few flowering vines. We planned to visit his nearby uncle, who had some hives, too, but the constant rain didn't cheer us up. I hoped to meet the uncle to inquire about the past riots, but I did not pursue it with Pak Ancah because he was just a little kid then.

Traditional Betang House

This Betang House is the creation of the Dayaknese Tribe that lived in Kalimantan inland with a group concept of life. At that time, Dayak tribes stayed inland of Kalimantan and spent their life in a group, together in a house called *'Rumah Betang'* (Betang traditional house, see Figure 66).

Betang has unique points that can be seen from the elongated shape, and there is only a staircase and entrance into the Longhouse. The stair used to enter the Longhouse is called *'hejot'*. Betang has a built-in high position from the ground to avoid trouble things for residents of the Longhouse, including avoiding enemies that can come suddenly, wild animals, or floods that sometimes hit the Longhouse. Almost all Betang can be found on the outskirts of major rivers in Kalimantan.

Usually, Betang is a built-in rather big size with about 30–150 meters in length, 10–30 meters in width and the mast about 3–5 meters in height. Betang was built using ironwood (*Eusideroxylon zwageri*). This wood can stand for hundreds of years and is anti-termite.

Betang is a tribe house because it is led by a Pambakas Lewu (the tribe leader). 100-150 people commonly inhabit Betang House; you may imagine how crowded inside. Inside, you will meet several rooms inhabited by each family.

On the front of the house is a hall to receive guests and the customary meeting place. Usually, here you will meet a *sapundu*. *Sapundu* is a statue or totem, which is generally formed human-shaped. *Sapundu* functioned as a place to tie the animals to be sacrificed for the ceremonial procession. Sometimes, there are also *patahu* in the front yard of the Betang that are used for worship houses.

On the back of the Longhouse, you will find a small hall called *tukau* used as a warehouse for storing agricultural equipment, such as *lisung* or *halu*. In Betang, a room is also used as a weapon storage area, called *bawong*. On the front or back part of Betang, there is also *Sandung*. *Sandung* is a storage place to save the bones of a family who had died and underwent the *tiwah* ceremony. The Dayaks of interior Kalimantan practised this form of ancestor veneration.

Chapter 20

Miniature Landscape and Meliponiculture

Landscape miniaturization is a Far East design philosophy where the garden comprises a mixture of trees, plants, hardscaping and small accessories, which are combined to create a lasting landscape (Manuela 2015). The art form, in essence, allows for appreciation of horticultural aesthetic values in a condensed scenario, usually onto a tray, though not larger than one that would need more than two people to move manually.

The origin of the terms used in presenting the evolution of miniature landscapes (adapted after Baran 2013, Baran 2010, Jonker et al. 2014, DuCane 1920, Kawamoto & Kurihara 1963, Terry Cheng 2009, Rahmann & Jonas 2009)

Chinese	Japanese	Literal translation	Description
pun or p'un, pen or p'en	bon	tray, pot	Tray or pot
	bonseki	tray stone	Landscape on a tray with stone and sand
pun-ching/ penjing	bonkei	tray landscape	Landscape in a container
pun-sai/ pun-tsai	bonsai	tray plant	Dwarf potted tree; artistic pot plant; table culture plant
	hachi-niwa/ hako-niwa	box garden	Japanese miniature garden; dish-garden
	saikei	planted landscape	Miniature living landscape (often using trees not yet developed enough to stand alone as bonsai and without as many formal rules)
	tsubo-Niwa	small gardens and courtyards	Courtyard garden or pocket garden

From the historical point of view, it is considered that the Japanese had experienced this form of cultivation by the end of the 13th century when the Chinese traders introduced these miniature trees to

Japan. Although the oldest representations of bonsai in Japan date back to 1300 AD, the bonsai may have been introduced in Japan by Buddhist monks before this period (10th -11th century).

The minimalism in Japanese art lies in miniaturising trees (bonsai) and creating miniature gardens (saikei and Hachi-Niwa) that evoke natural landscapes, combining miniature trees with soil and rocks, water and vegetation in a single container.

Saikei (planted landscape) is the art of creating miniature landscapes, reminding the viewer of a natural location by topography and using materials and species. *Saikei* focuses on evoking a natural landscape rather than individual tree appearance, as outlined in *bonsai*. In saikei, herbaceous plants (ground-covering plants, wild plants) evoke the landscape. Short ceramic trays are used for saikei, where vegetation and rocks are arranged.

Hachi-Niwa (box garden, miniature Japanese garden) are 'lilliputian' landscapes that include items such as a mountain, cliff or hill; a water pool with an island, a waterfall or a stream with a bridge; alleys, gates, a house; trees and shrubs, small plants, figurines & etc.

Tsubo Niwa is a courtyard garden or pocket garden in Japanese. One *tsubo* is equally 3.3 square meters in Japanese measurement. In the Heian Era, *tsubo* was an area surrounded by fences and buildings with different landscapes for viewing. The first courtyard gardens were designed and made between houses and storage in the 15th century. The main elements to design a *Tsubo Niwa* are trees, stepping stones, rocks, flowers, grass and water features (Terry, 2009).

One of the reasons why *bonsai* and miniature landscapes remain so popular in Japan is the lack of space; gardens often need to be brought indoors or addressed as miniature landscapes in a small courtyard, balcony, or container, evoking larger landscapes.

Like indigenous architecture, vernacular architecture would include indigenous landscapes and regional beescape. Part 5 ends with a gallery of illustrations of miniature landscapes with beehives and racks for ideas the beekeeper can choose from to fulfil the bees' foraging needs and reduce crawling predators.

Figure 179 Assorted Bonsai Models with Bee hives

Figure 180 Miniature Landscape models with Bee hives

Figure 181 Display Racks for Miniature Landscape Trays in Meliponiculture

Chapter 21

Meliponine Entomotourism

Introduction

Entomotourism is a subset of wildlife tourism, and it describes the pursuit of specific insects and their encounters in controlled settings. They may relate to an SB repository or sanctuary for Conservation, and educational strategies are used to increase public awareness of insects by emphasizing their ecological and socio-economic roles. Visitors may experience bee culturing techniques, honey harvesting, and honey-tasting activities. These tourist activities may provide funds for management and scientific research. On off days, researchers may conduct scientific experiments and analyses. Would you like to contribute something to this?

The Malaysia Genome & Vaccine Institute (MGVI) Meliponine repository[96] has provided an interactive conservatory that helps keep visitors connected and updated on the latest news from the entomological world. Recognizing and integrating insects into recreational activities in both young and old generations create a positive cycle of change where insects are tolerated and respected for their utility, diversity and adaptability.

Millions worldwide visit insectariums and butterfly pavilions, pollinator parks and beekeeping museums. Tourism that provides direct contact with wildlife creates a 'strong and positive educational note to visitors. They secure long-term conservation of national treasures and wildlife habitats.

Behaviours towards insects come from cultural, biological, and physiological factors. Entomotourism provides employment opportunities, revenue through arts and crafts sales, and regional economic diversification strategies for the local population. It also provides a value of the education and knowledge of insects.

Introducing integrated organic pest management while protecting bees into gardening practices, joining an insect association, participating in citizen science projects on insects, conducting backyard safaris or building an insect hotel, butterfly garden or dragonfly pond on the property are some of the helpful ways to change behaviours in the Entomotourism industry while making a positive impact. Protecting these bees and monitoring their habitat, nest hosts and appropriating the proper housing is key!

---

[96] A depository of different species of stingless bee colonies completes in their natural hive logs with their extension hive boxes attached on top. The main purpose is for the researchers to excess the hives for honey or pot pollen samples and insect specimens in their DNA and NMR analyses.

Native Beekeeping and cultural support for an indigenous village will support the local biodiversity, the community's resilience, education and protection of cultural heritage, food and economic security.

Integrating Vernacular Architecture

Most of the ASEAN countries, including India, have a cultural or heritage site as a Tourist destination. These Cultural villages are tourist hotspots and usually include replicas or imitation miniatures of the regional heritage and indigenous folk architecture. The miniature models may vary according to the available land acreage. Smaller areas may have miniature-scaled models showcased in museums. Such sites are in themselves tourist attractions. An added allure can be miniature-scaled model houses made as beehives. (See Prelude of Volume 1)

These indigenous or ethnic identity beehives explode with Insect Tourism potential. A vernacular structure with an ethnic identity or heritage significance adds a flare of Tourist attraction and, therefore, would gain much Entomotourism potential.

Figure 182 Sagada House in Tublay, Benguet, Central Luzon.

An example of beekeeping among indigenous people is the Sagada House in Tublay, Benguet, Central Luzon, Philippines. Bernard Anciado and Leo Kimbungan contributed these images in traditional garb as the tour guide in this instance. It's amazing how they preserved this 400-year-old last traditional house in Sagada.

There are other examples in North Borneo among some Kadazan beekeepers who keep bees in bamboo inter nodes and place them at their longhouses. Bruneians regularly visit these sites at the Kundasang Market, a favourite tourist spot. Kundasang is an agrarian hill town in Sabah, where villagers spread their daily produce and wares to passing tourists. It is situated in the interior highlands and en route to Mt. Kinabalu. It provides a panoramic view of the surroundings, and one may glimpse the unique

Kinabalu peak. Villagers used to hunt stingless bee nests in bamboo groves, cut out the internodes, and hang them for sale beside their stalls. They sold them for RM 70 to 80 per nest about a decade ago.

Figure 183 Kundasang Market town in Sabah, North Borneo. Bamboo hives are on display and for sale to tourists.

Tourists who are not beekeepers buy them for a one-time harvest of honey. Local and Brunei beekeepers purchase for propagation.

When we visited as tourists in Chiang Mai, North Thailand, we were directed to the Chiang Mai Dept. of Agricultural Extension Station, where a Giant replica of a Bee greeted us. The area is a large spread

Figure 184 Chiang Mai Dept. of Agricultural Extension Station visitor centre.

with both stingless bees and honeybee hives. They have a visitor's centre where products are displayed. Beside the building is a Shed/pavilion with Apiculture and Meliponiculture information notice boards.

In Indonesia, Bee farms in South Sumatra accommodate tourists attracted to their ornately carved bee housing with models of various indigenous architecture of different cultures in Indonesia.

Figure 185 Location of the Al Qorni qorni farm of Kalianda, Lampung, S. Sumatra. Hive box made with glass sides for observation.

They even have a couple of Glass Hives for viewing the internal nest structure (beats every time having to open for visitors wanting to view inside the hive).

In Aceh, North Sumatra, a beekeeper, Omjol, makes the traditional Aceh model house from Angsana wood measuring 50cm by 60cm. He keeps the overall design as a replica of his own traditional home.

Figure 186 A bee hive box in the design of a miniature model of a traditional Aceh house.

This stingless bee-keeper has operated this hive for 2 years without any snags. The roof is removable for inspection and harvesting activities of the hive. This Meliponary is located in Pereulak, East Aceh Regency, Aceh.

Meliponini Tourism in the Jungles of Kalimantan

Figure 187 Pusat Pendidikan & Pelatihan Lebah Tanpa Sengat Kalimantan Barat (PUSDIKLAT) West Kalimantan Stingless Bee Education & Training Center.

Stingless bee associations or co-operatives' collective show farms.

An example show farm by an association is Darul Naim, Entrepreneur of Stingless Bee Society (DRONESS) in the NE state of Kelantan, Peninsula Malaysia.

Figure 188 The DRONESS show farm in Kelantan. Photos by Dr. Orawan Duangphakdee

The images are from a technical tour during the recent conference International Stingless Bee Honey Standard: APIMONDIA Asia Initiative Meeting at USM (AAIM@USM) in Conjunction with The 5th International Conference On The Medicinal Use of Honey (V-ICMUH) and "World Bee Day Celebration on 18th May 2023" at University Sains Malaysia, Health Campus, Kubang Kerian, Kelantan.

Insect Museum in Brunei

Tasek Merimbun is Brunei Darussalam's only ASEAN Heritage Park since 29 November 1984. One of the attractions at Tasek Merimbun Heritage Park is the diverse bird species that inhabit the surface of the lake and the natural surroundings around it.

The complex of Tasek Merimbun was officially opened to the public in 2000. The complex comprises *Balai Pameran* / Exhibition Hall, *Balai Serbaguna* / Multi-purpose Hall (Balai Purun), *Makmal* / Laboratory, and *Taman Kulimambang* / Butterfly Garden (opened in 2012).

Figure 189 Specimens from Kuala Belalong are now placed in the Tasik Merimbun Museum. The display includes the actual log nest with the nest entrance intact and an illustration of the bee.

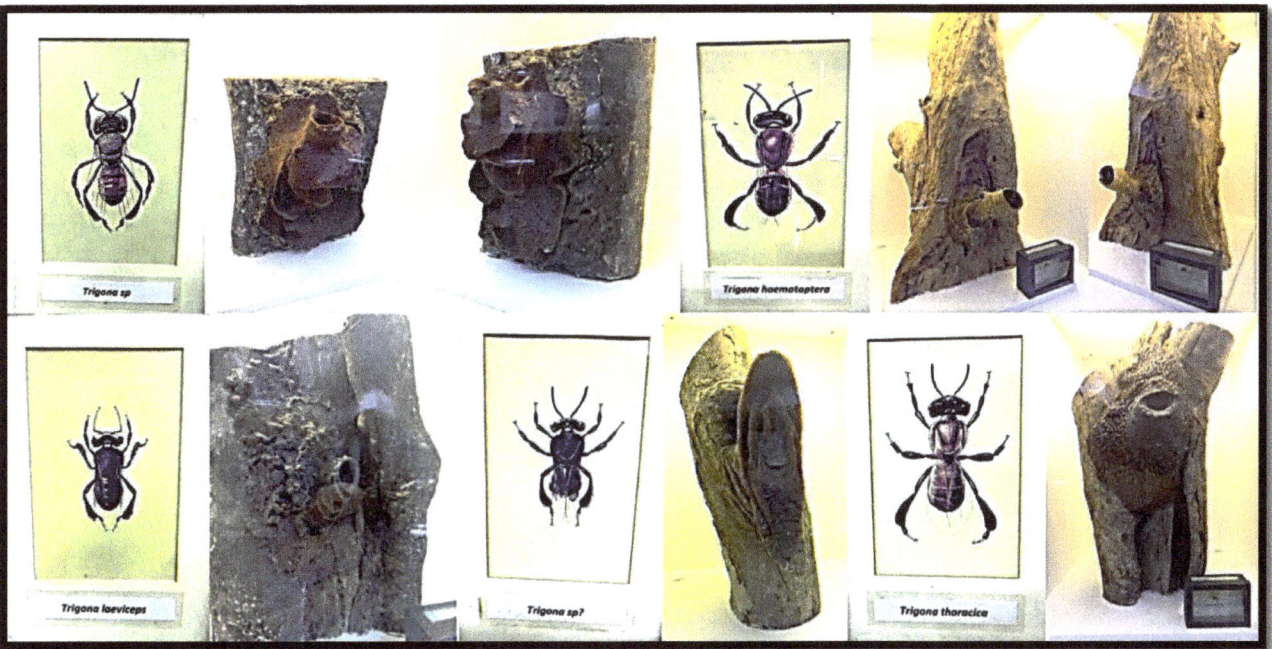

Figure 190 Here are some examples of kelulut exhibits at the Brunei Museum. It was collected by Dr. Roubik in 1976.

References for Entomotourism

Gao Le et al. (2021). *Entotourism potential in Sabah, Malaysia: A Tourists' perspective.* Retrieved from Cogent Social Sciences 7:1, 1914950: https://doi.org/10.1080/23311886.2021.1914950

Gao Le et al. (2021). Factors of Tourists' Perspective and Awareness of Insects Based on Demographics Distribution in The Tropical Ecosystem. *Serangga 2021, 26(2): 361-385.*, http://journalarticle.ukm.my/18872/1/48778-159572-1-PB.pdf.

Ismail, N., et al. (2014). Cultural Heritage Tourism in Malaysia: Issues and Challenges. *SHS Web of Conferences 12, 010 5 (2014)* (p. http://dx.doi.org/10.1051/shsconf/20141201059). Penang: EDP Sciences, 2014.

Kanaujia, A., et al. (2022). Insect Tourism. In Omkar(ed.), *Insects as Service Providers,* (pp. 283–305). Lucknow, Uttar Pradesh, India https://doi.org/10.1007/978-981-19-3406-3_10: Springer Nature Singapore Pte Ltd.2023.

Lemelin, R. (2012). The Management of Insects in Recreation and Tourism. In R. L. (Ed.), *The Management of Insects in Recreation and Tourism (p. Iii).* Cambridge: Cambridge University Press.

Lemelin, R. H. (2015). *The Evolution and Diversification of Entomotourism. In Animals and Tourism.* Retrieved from Fanimal - Recreational Fringe to Mainstream Leisure:: https://fanimal.online/wp-content/uploads/2021/10/Insect-Tourism-Entomo-tourism.pdf

Lundin, E. (2021.). Multispecies Interactions in Tourism: An ecofeminist exploration of tourist-insect encounters. *Digitala Vetenskapliga Arkivet*, https://www.diva-portal.org/smash/record.jsf?pid=diva2%3A1564932&dswid=7949.

Suprapti, A., Et A; (2019). The Spatial Concepts of Cultural Heritage Village Toward a Tourism Development: Case Study of Kadilangu Demak Indonesia. *Journal of Architecture and Urbanism Volume 43 Issue 1: 36–46*, https://doi.org/10.3846/jau.2019.6057.

Zakaria, M.Z., et al. (2020). Tourists' perceptions of insects as the determinants of insect conservation through entomological ecotourism. *Journal of Tropical Biology and Conservation (Malaysia), 17 (1), p. 79-95*, http://agris.upm.edu.my:8080/dspace/handle/0/21508.

Appendix - Glossary of roof types

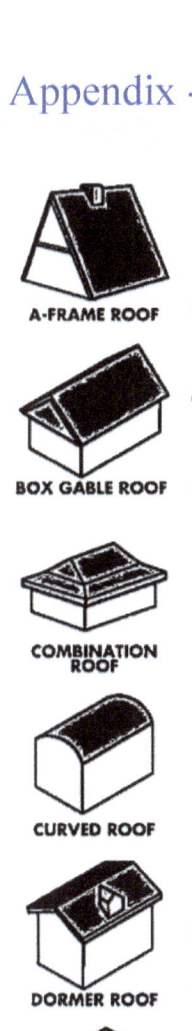
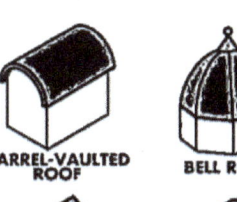

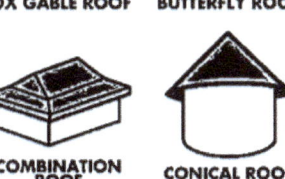

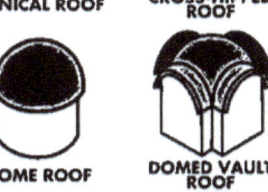
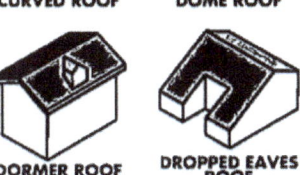

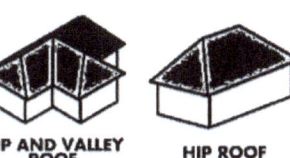
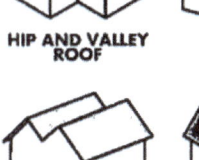
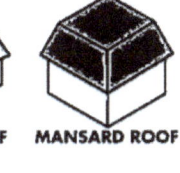
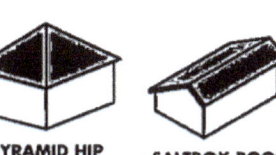
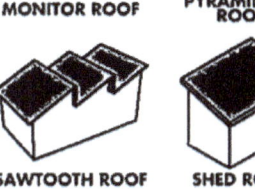

- **Table of Contents**
  - 1. A-Frame Roof
  - 2. Barrel Vaulted Roof
  - 3. Bell Roof
  - 4. Box Gable Roof
  - 5. Butterfly Roof
  - 6. Clerestory Roof
  - 7. Combination Roof
  - 8. Conical Roof
  - 9. Cross-Hipped Roof
  - 10. Curved Roof
  - 11. Dome Roof
  - 12. Domed Vault Roof
  - 13. Dormer Roof
  - 14. Dropped Eaves Roof
  - 15. Dutch Gable Roof
  - 16. Flat Roof
  - 17. Gable Roof
  - 18. Gambrel Roof
  - 19. Hexagonal Roof
  - 20. Hip and Valley Roof
  - 21. Hip Roof
  - 22. Jerkinhead Roof
  - 23. M-Shaped Roof
  - 24. Mansard Roof
  - 25. Monitor Roof
  - 26. Pyramid Hip Roof
  - 27. Saltbox Roof
  - 28. Sawtooth Roof
  - 29. Shed Roof
  - 30. Skillion Roof

Every roof is designed with a specific set of benefits in mind. These benefits are often related to the climate and environment in the area where they're built. While some roof types are much more common than others, every roof type has an ideal purpose and use, whether it's because of its simplicity, cost-effectiveness, or functionality.

1. A-Frame Roof

An A-frame roof has a steeply angled (roofline) that typically functions as both the roof and two of the four exterior walls. The roof will usually begin at or near the foundation and join at the top in a way that resembles the letter "A".

The A-frame roof is a simple and less expensive roof design because the roof serves as both a roof and a wall. A-frame roofs can be found worldwide, notably in Europe, North America, China, and the South Pacific islands.

2. Barrel Vaulted Roof

Barrel vaulted roofs have a curved roof that resembles a barrel cut in half. Barrel roofs are designed with a similar purpose to a dome roof, with the primary advantage over dome roofs being that they can cover longer, rectangular buildings.

A barrel vault is the simplest type of vaulted roof, with a curved, often semi-circular cross-section with a semi-cylindrical appearance that can span long distances. In cases where windows are installed along the roof of a building, barrel-vaulted ceilings can be beneficial in allowing the maximum amount of light into a building, and they also allow for additional height that would not otherwise be available in other roof types.

3. Bell Roof

As its name suggests, a bell roof is a roof resembling the shape of a bell. This roofing type often comes in different forms, such as round, square, and multi-sided. Because of its classic appearance, a bell roof is typically found on various vintage-looking structures, including Colonial-era churches, schools, and historic homes.

4. Box Gable Roof

Box gable roofs are distinguished by their triangular extensions at each face of the house, with the roof boxed at the end. This roofing type looks similar to a regular gable roof but with the triangular extensions closed off instead of being left open. Box gable roofs have a more pronounced triangular shape than regular gable roofs.

5. Butterfly Roof

A butterfly roof generally comprises two tandem pieces that meet midway and are angled in a V-shape. This type gives the effect of a butterfly's wings in flight when seen from the exterior, making it a good choice for modern and contemporary homes. Not all butterfly roofs need their two pieces to meet midway physically; they only need them to slope inward at a midpoint.

6. Clerestory Roof

A clerestory roof has two sloping sides joined by a short, vertical wall. The slope typically falls outward, meaning the peak is near the roof's middle. Its most prominent feature is a row of even, horizontal windows (or one long, continuous window) on the exposed face of the vertical wall.

This type of roof can be symmetrical or asymmetrical, resembling a skillion roof, and has a hipped or gable design.

7. Combination Roof

Combination roofs are, quite literally, a blend of various roofing styles based on the building's theme and environment. There is no limit on how many types this roof can incorporate or which ones can be combined. For instance, it could be multiple gable roofs with a hip roof over the front porch or a hip roof paired with a clerestory roof for a distinct look.

Due to their unique architectural appeal, combination roofs are well-suited to modern and contemporary homes but are also seen in some vernacular or ethnic dwellings.

8. Conical Roof

Also known as a cone roof, witch's hat, or turret roof, a conical roof is round on a flat pane and rises to a point, forming a cone shape. Like other common roofing styles such as gable and hip roofs, conical roofs also have roof rafters and support columns, although, most of the time, they are shaped unusually and cut at different angles to match the cone form. Some conical roofs project out of the wall to form eaves around the usually circular or multisided structure.

9. Cross-Hipped Roof

Cross-hipped roofs (or cross-hip) are some of the most popular variations of the standard hip roof design, which can be thought of as two hip roofs joined together at a right angle. These are most often laid over buildings in a T or L shape. Cross-hipped roofs come with two intersecting hip sections that run perpendicularly. The seam forms the cross-hipped roof, and the two sections meet at the end, forming a valley.

These roofs are great for structures with a more complex layout rather than a usual rectangular or square and can stand up to rain, snow, and high winds incredibly well.

10. Curved Roof

Often used as an alternative to a flat roofing system, a curved roof is usually attached to a taller exterior wall. It forms an arch that can go from a low slope to a more rounded peak, allowing for water runoff while adding value to architectural interest. A curved roof can be used for various home features, such as an addition or wing, an arched entrance, or an entire building. The arch may vary from hyperbolic to parabolic shape. Due to its naturally bent shape, this roofing type often requires a flexible rafter material.

11. Dome Roof

Dome roofs are polygonal and characterized by an inverted bowl shape. While relatively costlier than other alternatives, dome roofs are extremely durable and perfect for specific home features like gazebos or cupolas. When planned and designed properly, they can also be used as the main roof of a building. The cross-section may be semi-circular, varying to hyperbolic.

12. Domed Vault Roof

This type is a variation of the dome roof but with a few differences. The domed vault has self-supporting arches, adding drama and visual interest to the structure. Domed vault roofs shed water easily, which is why they are great for areas with a lot of precipitation. Most architects, however, choose them primarily due to their aesthetic appeal.

13. Dormer Roof

The roof that covers this window is called a dormer roof. A dormer roof is not an entirely separate type of roof, but it takes its name from a dormer or a windowed structure that projects vertically from a sloping roof. Dormers are used to make attic spaces liveable, which would otherwise be dark and cramped.

Dormers can also significantly enhance a house's aesthetics, often functioning as the "eyes" of the home. They also come in various forms, including gable dormers, hip roof dormers, flat roof dormers, and wall dormers. Most of the time, the materials used on a dormer roof are similar to those used on the main roof.

14. Dropped Eaves Roof

Dropped eaves roofs have two slopes that meet at the centre, much like a saltbox roof. Their slanting side on the front has steep eaves, hence their name. This slope results in rooms with slanted ceilings, increasing its appeal among many homeowners. At the same time, it also helps direct water efficiently to the gutters and downspouts, preventing standing water that could damage the roof materials.

Dropped eaves roofs are often used as a canopy for entrance stairways in some vernacular designs, especially those on stilts.

15. Dutch Gable Roof

A Dutch gable combines a hip and gable roof, using two roofs in one dwelling. It uses a hip roof as a "base" over which a smaller gable roof sits in the centre. As a result, it increases attic space and complicates bracing the structure for storms and wind uplift.

Since a standard hip roof makes the attic look and feel cramped, the gable roof offers a comparatively more spacious room inside. Meanwhile, the hip structure provides more durability and strength, resulting in better weather resistance for the entire roofing system. A similar structure can be found in East Asian hip-and-gable roofs.

16. Flat Roof

While they look flat from the outside, flat roofs are not exactly flat—they have a slight incline to prevent water from collecting on the top. The National Roofing Contractors Association defines a roof with a slope of 3-in-12 or less as a "flat roof."

17. Gable Roof

Also known as peaked or pitched roofs, gable roofs are some of the most common roofing types seen in many homes worldwide. They are easily recognized by their standard triangular or inverted "V" shape. Many homeowners prefer them mainly because they shed water and snow easily, allow for more attic ventilation, and are cheaper and easier to build than other roofing types. However, one of the downsides of a gable roof is that it's prone to storm damage, especially if there is a significant overhang.

18. Gambrel Roof

A gambrel roof is like a gable roof, but only if you add another slope to its lower edges. They are also known as "barn roofs" because they are often seen on top of barns, farmhouses, and log cabins. However, they can also be used in many traditional homes, giving the Dutch Colonial influence. Due to their shape, they help provide more storage within a building. The simpler construction also allows the gambrel roof to use only two roof beams, making the roof vulnerable to immense stress in strong winds.

19. Hexagonal Roof

Hexagonal roofs are, essentially, roofs with six sides that slope downward. Due to their unique and polygonal structure, these roofs are unsuitable for every residential or commercial building. They are mainly used to improve a structure's aesthetic appeal rather than utility. Pavilions, cabanas, and gazebos are some of the most common structures that use hexagonal roofs.

20. Hip and Valley Roof

This type of roof has both hips and valleys. Compared with roofs with a single shape to cover a house, this roofing type has numerous dips and peaks, adding a dramatic touch and accent to any structure. A hip and valley roof is standard among many Colonial structures and pre-20th century houses. Nowadays, it is also being used in various modern homes.

21. Hip Roof

Unlike regular gable roofs that do not have sloping structures on two sides of the building, hip roofs have slopes on all four sides. These sides are all equal and join at the top, forming a ridge. This design makes them more structurally stable than gable roofs.

Due to the inward slope on all sides, hip roofs are often sturdier and more durable than their gable counterpart. They can be covered with roofing material and modified to include dormers or crow's nests. They are great for areas prone to high winds and snow, as the slant of their slopes lets the snow slide off easily.

22. Jerkinhead Roof

Jerkinhead roofs combine two of the most popular roofing types: gable and hip. However, unlike Dutch gable roof, which is also a fusion of both types, jerkinhead roofs are huge gables topped with a flattened, clipped edge, similar to a hip roof.

Jerkinhead roofs offer more attic space and greater wind stability than a standard hip roof.

23. M-Shaped Roof

An M-shaped roof is a double-gable roof. The roof rests on two bearing walls with two sloped sides that meet at the centre, forming an "M" shape. Unlike standard roofing types with only gutters around the edges, M-shape roofs have a central gutter system between the two pitches, preventing snow and water from building up during the winter.

24. Mansard Roof

Like the gambrel roof, mansard roofs have two different slopes on the sides of each roof. But while the gambrel only has two sides, a mansard roof has four, which makes it an analogue to a hip roof (as the gambrel is to a gable). In a mansard roof, the lower slope is much steeper than the upper, and all sides can be flat or curved, depending on the building's architectural style. The lower, steeper slope can be used as additional floor space, known as a garret, and can be punctuated with dormers.

25. Monitor Roof

A monitor roof is a raised superstructure that runs along the ridge of a double-pitched roof. With its long sides, property owners can install clerestory windows or louvres to help boost lighting and air circulation within a building. Clerestories built this way can also accommodate vents. Monitor roofs that run the full length of a building are uncommon on residential properties; instead, they are commonly found in warehouses, barns, and factories, where maximizing additional light from windows at a greater height can be useful.

26. Pyramid Hip Roof

As its name implies, a pyramid hip roof takes the shape of a pyramid and is constructed on top of a square or rectangular base. It features four triangular sides like a standard hip roof, but instead of forming a ridge at the centre, the sides of a pyramid hip roof converge at a single point, which means it doesn't have the ridge as does a standard hip roof.

This type of roof is suitable for small structures, such as cabins, or small sections of the home, like garages and pool houses.

27. Saltbox Roof

A saltbox roof is known for its asymmetrical shape, which resembles antique wooden saltboxes of the 1700s. A saltbox roof features two slopes of varying lengths, with one much longer than the other. An easier visual representation would be looking at it like a gable roof with one side shorter. The slope may differ on each side, and the longer side may even reach low enough to allow ground-creeping vegetation to cover a "green roof".

Buildings with this roofing type usually have two stories on one side and one on the opposite. Because of its sloped structure, water runs off easily on this roof, making it ideal for areas that receive heavy rains.

28. Sawtooth Roof

A sawtooth roof comprises a series of ridges with double pitches on each side. They have numerous parallel planes that resemble a saw's teeth, with one slope looking steeper than the other one. Windows are often installed in the vertical spaces of the roof, allowing plenty of natural light to pass through. Sawtooth roofs are commonly used on large greenhouses and industrial buildings but can also be seen in many modern houses today.

29. Shed Roof

Shed roofs have a single slope that inclines at a certain angle. This roofing type is common among many contemporary homes primarily because it offers more interior space than the popular gable and hip roofs while maintaining a simple, sophisticated appearance. Other structures that use shed roofs include animal sheds, outhouses, and storage barns.

30. Skillion Roof

Skillion roofs are similar to shed roofs because they both have a single flat surface with a steep and noticeable pitch. They are an excellent choice for buildings in rainy and snowy regions, as their steep slope allows water and snow to shed easily. However, unlike the latter, skillion roofs can come in numerous planes—for example, the butterfly roof, with an inverted "V" shape, is known as a variation of the skillion roof.

Modified from https://retipster.com/roofs/ by Seth Williams

List of Figures

Figure 1 The superimposed green represents areas with the .... - 11 -
Figure 2 Inspired by the King Faisal Mosque in Islamabad, Pakistan. .... - 12 -
Figure 3 Latest earthquake in Pakistan 2022. Source:
https://earthquake.usgs.gov/earthquakes/eventpage/us7000i57m/executive .... - 12 -
Figure 4 Recent Earthquakes in Pakistan. Afghanistan and Tajikistan Region as of 22 Niv, 2022 .... - 13 -
Figure 5 Distribution of Meliponiculture activities in the Indian Subcontinent .... - 14 -
Figure 6 Type localities for the species of stingless bees .... - 15 -
Figure 7 Vernacular Architecture in the Indian Subcontinent .... - 16 -
Figure 8 A Salvaged South Indian House in the Heritage Village in Manipal, Karnataka. Drawn by AHJ ...... - 17 -
Figure 9 a) & b) Madhusree Bee Farm c) Stingless bee trapping d) & e) PVC pipe housing for bees f) Painted wooden box hives .... - 20 -
Figure 10 Andaman and Nicobar Islands village hut .... - 21 -
Figure 11 Munda_House_at_State_Tribal_Fair-020__Bhubaneswar .... - 22 -
Figure 12 Kisan tribal house at the 2020 Odisha State Tribal Fair, Bhubaneswar .... - 22 -
Figure 13 Oraon House at 'State Tribal Fair-2020' Bhubaneswar, India .... - 22 -
Figure 14 Santal house at 2020 Odisha Tribal Fair, Bhubaneswar .... - 23 -
Figure 15 Left: A Gadaba hut, Koraput, Odisha; Middle: A Village complex in Andhra Pradesh; Roght: A typical house of tribals of the Gadaba community from Odisha. .... - 23 -
Figure 16 Lanjia Sora house at Odisha State Tribal Fair, Bhubaneswar .... - 23 -
Figure 17 A typical house of tribals of the Bathudi community from Odisha. .... - 24 -
Figure 18 A traditional Khond house. .... - 24 -
Figure 19 A traditional Adi hut (interestingly, it looks similar to the Ibaloi House in The Philippines) .... - 25 -
Figure 20 Singpho dwelling in Arunachal Pradesh .... - 26 -
Figure 21 Rendering of T. gressitti (Sakagami 1978) by AHJ .... - 26 -
Figure 22 Katkari, also called Kathodi dwelling http://www.rainforestinfo.org.au/projects/india/Katkari.htm - 27 -
Figure 23 This typical home belongs to the Tadvi Bhils in Maharashtras Satpuda region .... - 27 -
Figure 24 Sawantwadi dwelling .... - 27 -
Figure 25 Mardhari village vegetables like pumpkin are grown on the roofs of homes .... - 28 -
Figure 26 "Dhīmar or fisherman's hut." from The Tribes and Castes of the Central Provinces of India Volume II Author: R. V. Russell .... - 28 -
Figure 27 Figure 18 Rendering of T. ruficornis (Smith in Horne & Smith 1870) by AHJ .... - 28 -
Figure 28 Vernacular Architecture of West Bengal- Zonal Geography Classification According to Climatic Conditions .... - 29 -
Figure 29 Do-chala roof- Left: Damodar temple of Siuri in Birbhum district; Right: Jor Bangla Temple, Bishnupur with a curved Do-chala style roof .... - 30 -
Figure 30 Char-chala style roof; Left: Palpara Temple – Nadia 2011; Right: Palace in Deeg, Rajasthan (in 1750) .... - 30 -
Figure 31 At-chala roof – Atchala and Pancha Shiva Mandir in Pathra; Right: Antpur Radhagovindjiu Temple; - 30 -
Figure 32 Ratna roof – Left: Ram Chandra Temple, Guptipara, Hooghly district; Right: Radha Govinda Temple Bishnupur Bankura .... - 31 -
Figure 33 Ek-ratna roof – Ekratna Temple of Gopinath, Khard Radhakantapur .... - 31 -
Figure 34 Dalan roof – Left: Dalan Temple inside Kaviraj Bari at Mankar situated in Purba Bardhaman district; Right: A 17th century haveli (a mansion) in Old Dhaka .... - 31 -
Figure 35 Rasmancha Roof - Hindu ritual platform, the Rasmancha, Bishnupur, c. 1600 .... - 32 -

Figure 36 Left: Antpur Chandimandap; Middle: An 18th-century Rajput painting by the artist Nihâl Chand; Right: The Naulakha Pavilion in Lahore Fort, Pakistan, features a Do-Chala roof originating in Bengal. ...... - 32 -
Figure 37 Left: A multi-domed Sultanate-era mosque; Middle: Mughal-era domes in Murshidabad; Right: Choto Sona Mosque (around 1500) .... - 32 -
Figure 38 Imperial Delhi Moti Masjid .... - 33 -
Figure 39 Left: Bungalow in Bengal, Bangladesh; Right: Momin Mosque since 1920 .... - 33 -
Figure 40 The Himalayan states cover Nepal, Bhutan, Assam, Nagaland and North Bangladesh. .... - 34 -
Figure 41 Left: Plains Tiwa's Hut; Right: A Hill Tiwa house. .... - 34 -
Figure 42 The royal seat of Khyrim at Smit .... - 35 -
Figure 43 Tetragonula bengalensis (Cameron 1897) .... - 35 -
Figure 44 Lepidotrigona arcifera Photo Claus Rasmussen .... - 36 -
Figure 45 Village dwellings in The Upper Dzongu forest block in Sikkim .... - 36 -
Figure 46 Similarities in the 11th-century Nordic battle preparation camp and the construction of the Aguarano home in Peru. .... - 37 -
Figure 47 The ethnic tribes in Nagaland, however, have a unique construction. .... - 37 -
Figure 48 Beehouse in the design of the Sema Naga house. .... - 38 -
Figure 49 Tribal structures in Nagaland structures are very similar to Viking houses. .... - 38 -
Figure 50 Meliponiculture in Nagaland by Bode Shuya .... - 39 -
Figure 51 Top left: Local Thenzawl architecture and Traditional structures in Mizoram, NE India. .... - 40 -
Figure 52 Distribution of Meliponiculture in the Indian Subcontinent .... - 41 -
Figure 53 Inspired by the Typical Nepali "Goal Ghar" or roundhouse .... - 42 -
Figure 54 Drukpa Namgyal Lives in Damphu, Chirang, Bhutan .... - 42 -
Figure 55 Vernacular architecture in Sri Lanka and South India. .... - 43 -
Figure 56 Tetragonula iridipennis (Smith 1854) .... - 44 -
Figure 57 Tetragonula praeterita (Walker 1860) .... - 44 -
Figure 58 Patthirippua at Sri Dalada Maligawa Kandy, Sri Lanka (The Temple of the Tooth) .... - 44 -
Figure 59 Left: Wanniyala-Aetto Village Right: Veddahs (wild men), Ceylon. .... - 45 -
Figure 60 Mud House in NW Sri Lanka .... - 46 -
Figure 61 The Mawlynnong house In the Meghalaya state in N E India - The Cleanest Village In Asia: 2022. .... - 46 -
Figure 62 Thailand upland-dwelling peoples (also known as "hill tribes") .... - 48 -
Figure 63 Distribution of three different peoples usually called "Sea Nomads" or "Sea Gypsies": Blue: Moken Orange: Orang Laut Green: Sama-Bajau .... - 50 -
Figure 64 Unique Architecture of Rong House .... - 51 -
Figure 65 Traditional Thai village homes with many similarities with the Malay village house design .... - 52 -
Figure 66 Two main types of Chofah: Pak Hong; Swan's tip (left) and Pak Khrut; Garuda's tip (right) .... - 52 -
Figure 67 Left: Oriental mythical figures that may adorn Indo-Chinese roofs; Right: Oriental influence on a Vietnamese house roof applied on a box hive with the rooster roof ornaments and upswept roof eave corners. .... - 53 -
Figure 68 Filipino roof ornament Sarimanok (Maranao) .... - 53 -
Figure 69 Comparing Rake and Fascia boards in Indo-China .... - 54 -
Figure 70 https://upload.wikimedia.org/wikipedia/commons/5/57/Mandalay_Palace.JPG .... - 55 -
Figure 71 Ancient Khmer Architecture influenced pediments, rake board facia and finials. .... - 55 -
Figure 72 Model of a Dai bamboo house. .... - 56 -
Figure 73 Left: The Palaung in the early 1900s; Right: Palaung village near Kyaukme, Myanmar (2017) ... - 56 -
Figure 75 The Sagwa Kayin House .... - 56 -
Figure 75 The Paw Kayin House .... - 56 -
Figure 76 The Bwe Kayin House .... - 57 -

Figure	Description	Page
Figure 77	Shan Tribal House Redrawn from Oranratmanee, R. (2018)	- 57 -
Figure 78	Stilt houses around Inle Lake, some with unique vernacular architecture	- 58 -
Figure 79	Tetragonula laeviceps (Smith, 1857)	- 59 -
Figure 80	Regional Myanmar Housing	- 61 -
Figure 81	Tetragonula pagdeni Schwarz	- 64 -
Figure 82	Top left: Nest entrance; Top Right: Internal nest structure; Bottom left and middle: Brood cells (Cluster–type nest) Bottom right: Storage cells	- 64 -
Figure 83	Tetragonula laeviceps Smith	- 65 -
Figure 84	Left: Nest entrance; Middle: Brood cells ; Right: Storage cells	- 65 -
Figure 85	Lepidotrigona terminata Smith & Nest entrance	- 65 -
Figure 86	*Left: Brood cells (Horizontal spiral brood comb); Middle: Storage cells; Right: Involucrum*	- 65 -
Figure 87	Lepidotrigona ventralis Schwarz & Nest entrance	- 66 -
Figure 88	Left: Brood cells; Middle: Involucrum; Right: Storage cells	- 66 -
Figure 92	Houses in the Central region have horn-like structures at the base of the gable	- 69 -
Figure 92	Northern houses have a "kalae" (V-shaped carving at the gable peak)	- 69 -
Figure 92	Houses from the Northeast region or Issan have a "Kaan" (bamboo organ).	- 69 -
Figure 92	Southern houses have an ornate carving around the rod at the apex of the gable	- 69 -
Figure 93	Thai Rake Boards	- 70 -
Figure 94	Typical Thai Finial	- 70 -
Figure 95	Regional Types of Thai House (Redrawn from Antarikananda, 2005)	- 71 -
Figure 96	Illustration of Tetragonula hirashimai	- 72 -
Figure 97	Tetragonula (Tetragonula) malaipanae Engel, Michener & Boontop 2017	- 73 -
Figure 98	Lepidotrigona satun collected in the Province of Sarun near the Thailand – Malaysia border	- 73 -
Figure 99	Assorted Bamboo huts for Bee friendly environment	- 74 -
Figure 100	Types of wall cladding	- 75 -
Figure 101	Dinner in Muslim Restaurant in Chiang Mai. From left: Dr. Bajaree Chuttong, Sangdao Banziger, Dr. Hans Banziger, AHJ and AHJ's wife Zaiton.	- 76 -
Figure 102	The Akha Tribal Hill slope House and a traditional house in Namo Nua Village, Laos.	- 76 -
Figure 103	A floating tomato garden on Inle Lake	- 76 -
Figure 104	Khmer traditional house in front of Angkor Wat taken between 1919 - 1926.	- 79 -
Figure 105	12th century-Bayon temple's bas relief	- 79 -
Figure 106	Pteas Pit house. Source: https://www.facebook.com/ASEANCambodia/photos	- 80 -
Figure 107	Pteas Rongdorl	- 80 -
Figure 108	Pteas Rongdeung or similarly a Cham House	- 81 -
Figure 109	The simplest Pteas Kontaing's roof	- 81 -
Figure 110	A type of traditional Khmer house known as Pteas Khmer	- 82 -
Figure 111	A type of Cambodian house that was built only for senior officials	- 82 -
Figure 112	Traditional Rest House Source: https://en.wikipedia.org/wiki/Traditional_Khmer_Housing#/media/File:Khmer_traditional_resthouse_in_1930.jpg	- 83 -
Figure 113	A traditional Khmer house in the mid-1800s taken by Emile Gsell.	- 83 -
Figure 114	Bee Gallery in Cambodia – Basic roof design is Pteas Khmer	- 84 -
Figure 115	Hobby Horse House (Phteah Chiahseh)	- 85 -
Figure 116	House with Grey Shutters (Phteah Betotvear Pnrbrapheh)	- 85 -
Figure 117	House of Many Roofs (Phteah Dambaul Chraen	- 85 -
Figure 118	original Khmer tall house or "phteah khpasa" is a stilt house or pile house.	- 86 -
Figure 119	The "Fishmongers' House" (phteah anaknesaeat)	- 86 -
Figure 120	The House with Horizontal Cladding (Phteah Kdab Phtek)	- 87 -

Figure 121 Traditional house in a settlement .... - 87 -
Figure 122 Example Traditional Village Houses .... - 88 -
Figure 123 Traditional waterside Khmer house .... - 89 -
Figure 124 Further developments using modern building materials .... - 89 -
Figure 125 The basics of the Cham people (Photo source:
https://commons.wikimedia.org/wiki/Category:Cham_people#/media/File:In_front_of_Ch%C4%83m_house.jpg .... - 90 -
Figure 127 An outdoor exhibit at the museum, consisting of an Ede dwelling from the Central Highland region. .... - 91 -
Figure 127 An exhibit at the Vietnam Museum of Ethnology in Ninh Thuan province. Source: http://www.vme.org.vn/en .... - 91 -
Figure 128 A modified Cham House turned Bee Gallery. .... - 91 -
Figure 129 Location of the PEKABA visit. .... - 92 -
Figure 130 Bee hive design inspired by an ancient Cham ruin
https://en.wikipedia.org/wiki/M%E1%BB%B9_S%C6%A1n .... - 93 -
Figure 131 Typical Kampong Cham Traditional Houses .... - 93 -
Figure 132 A Hani house in Vietnam Source:
https://en.wikipedia.org/wiki/Hani_people#/media/File:Nh%C3%A0_ng%C6%B0%E1%BB%9Di_H%C3%A0_Nh%C3%AC_(m%E1%BA%B7t_b%C3%AAn).jpg .... - 94 -
Figure 133 In places like Honghe, Yuanyang and Luchun, houses have mud walls (bricks, made from mud or clay) and thatched roofs supported by wooden pillars placed on stone foundations .... - 94 -
Figure 134 20111102-wiki com Akha Hut.JPG .... - 95 -
Figure 135 Green roofed Hani 'Mushroom House' .... - 95 -
Figure 136 a) b) & c) Meliponine hive station; d)storage pots in the bamboo hive; e) brood cells in the bamboo hive; f) & g) preparing to transfer the colony; i) transfer into a box. .... - 98 -
Figure 137 a) Log hive; b) nest entrance; c) brood cells clustered; d) High sugar content in honey; e) Provision of high-quality honey retains aroma, colour and taste after refrigeration; e) pot-pollen harvest .... - 99 -
Figure 138 Tuấn Vũ of National Association of Beekeeping [HỘI NUÔI ONG DÚ TOÀN QUỐC]
https://www.facebook.com/groups/2134975603386858/user/100002286322548 .... - 99 -
Figure 139 Stingless bee box hive .... - 101 -
Figure 140 a) b) c) assorted nest entrances d) Internal nest structure e) Tetragonula sp. f) Tetragonula queen g) established box hive h) brood cells, honey & pollen pots .... - 103 -
Figure 141 a) log hives newly placed b) log hives hung up c) Bamboo hive d) purpose-made wooden box e) manual harvesting f) box hive diagram g) Honey from box hive. .... - 104 -
Figure 142 Typical Tetragonula sp. Nest in a shed post. .... - 107 -
Figure 143 Small stingless bees make nests in village trees .... - 108 -
Figure 144 Nest entrances found in wooden House posts .... - 108 -
Figure 145 Location of Akha Tribal villages .... - 110 -
Figure 147 Inspired by the Karen Hill tribe traditional architecture. .... - 111 -
Figure 146 Akha Tribe in Chiang Rai province near the Thailand & Laos Border. .... - 111 -
Figure 148 Inspired by a Tai Dam (Black Tai) tribal house in Laos. .... - 112 -
Figure 149 A study of the architecture of the ancestor veneration practices prevailing in South China. ... - 113 -
Figure 150 South Chinese architectural influence in North Vietnam. https://www.alotrip.com/about-vietnam-culture/brief-history-vietnam-architecture .... - 113 -
Figure 151 Hokkien Chinese (Southern Min language originating from the Minnan region in Southeastern Mainland China) celebrated their 9th day of Chinese New Year last night. This celebration is rampant throughout Malaysia and Singapore .... - 114 -
Figure 152 Examples of a pyramidal roof on Box hives .... - 114 -

Figure 153 A Sinadumparan Ivatan house is one of Batanes Islands' oldest structures. The house is made of limestone and coral, and its roofing is cogon grass. .... - 115 -
Figure 154 Temperature and humidity recordings .... - 118 -
Figure 155 Thermal imaging .... - 119 -
Figure 156 Highest Recorded Temperatures in the ASEAN.. - 119 -
Figure 157 Windmills as his Meliponary theme by KT Chan of Alor Janggus, Kedah, Peninsular Malaysia.. - 123 -
Figure 158 Conceptually for yogic meditation, but looks like a good protective cover for box hives during typhoons.. - 124 -
Figure 159 An example covering all weather .... - 124 -
Figure 161 Anchor by footing and ground beam for maximum stability. .... - 125 -
Figure 160 Inspired by the bus shelter concept to shelter beehives .... - 125 -
Figure 162 Left: Mr. Nasiruddin chose this roofing style for his bee box hives in his Meliponary behind the Langkawi Cultural Centre: the ornately carved bee box; right: a proposed bee house to blend into the Langkawi landscape, the concrete stairway anchors the whole structure at high tides. .... - 126 -
Figure 163 With Aceh roots and Thai blended into the Traditional Malay House, these have grown into the typical Langkawi traditional homes. .... - 126 -
Figure 164 A typical fishing village row of houses... - 127 -
Figure 165 The stairway is usually receded into the building under the main roof structure................. - 127 -
Figure 166 This Governor's Mansion in Phuket City has the Malay Gable and Hips style roof with a partial Hexagonal central foyer-like section. .... - 128 -
Figure 167 The Istana Jahar Palace 19th century wooden building is now Kota Bharu Malaysia's Museum...... - 128 -
Figure 168 Istana Ulu (Early Palace of Perak) in Kuala Kangsar, Perak Darul Ridzuan, Malaysia, incorporates many ornamental elements of mansion designs in the Sino-Portuguese style. .... - 129 -
Figure 169 Vernacular fusion of the Bonnet roof with a 2nd pitch skirting. Bottom left inset: A Dutch doll bonnet. .... - 129 -
Figure 170 Inspired by the Orang Ulu tribe of Sarawak. .... - 131 -
Figure 171 Ornately carved bee box hive; Right: AHJ and Agus Suhartanto with his assistants. .... - 132 -
Figure 172 AHJ with Loloq, the pet owl... - 132 -
Figure 173 Left: Mandau hilts carved by Agus; Mid left: Totem poles at the museum; Mid right: a Dayak Belawing (clan identity pike); Right: Bee rack with traditional Kutai Dayak motifs. .... - 133 -
Figure 175 Gaddang Tribe of Solano .... - 137 -
Figure 174 The Ilongot or Ibilao house of Luzon and the Ilongot Tower .... - 137 -
Figure 176 An Aztec Golden Pavilion and a skull rack (from https://en.wikipedia.org/wiki/Aztecs#Aztec_culture) .... - 138 -
Figure 177 Image source https://en.wikipedia.org/wiki/Longhouse and https://en.wikipedia.org/wiki/Dayak_people .... - 139 -
Figure 178 AHJ with the family of Pak Ancah. Penggilingan Padi, Sampit .... - 139 -
Figure 179 Assorted Bonsai Models with Bee hives .... - 143 -
Figure 180 Miniature Landscape models with Bee hives .... - 144 -
Figure 181 Display Racks for Miniature Landscape Trays in Meliponiculture... - 145 -
Figure 182 Sagada House in Tublay, Benguet, Central Luzon. .... - 147 -
Figure 183 Kundasang Market town in Sabah, North Borneo. Bamboo hives are on display and for sale to tourists... - 148 -
Figure 184 Chiang Mai Dept. of Agricultural Extension Station visitor centre. .... - 148 -
Figure 185 Location of the Al Qorni qorni farm of Kalianda, Lampung, S. Sumatra. Hive box made with glass sides for observation. .... - 149 -

Figure 186 A bee hive box in the design of a miniature model of a traditional Aceh house. .... - 149 -

Figure 187 Pusat Pendidikan & Pelatihan Lebah Tanpa Sengat Kalimantan Barat (PUSDIKLAT) West Kalimantan Stingless Bee Education & Training Center. .... - 150 -

Figure 188 The DRONESS show farm in Kelantan. Photos by Dr. Orawan Duangphakdee .... - 150 -

Figure 189 Specimens from Kuala Belalong are now placed in the Tasik Merimbun Museum. The display includes the actual log nest with the nest entrance intact and an illustration of the bee. .... - 151 -

Figure 190 Here are some examples of kelulut exhibits at the Brunei Museum. It was collected by Dr. Roubik in 1976. .... - 151 -

Index

A

Aceh architecture, - 127 -
Acehnese, - 126 -
aesthetic appeal, - 156 -, - 157 -
A-Frame, - 153 -, - 154 -
Aguarano, - 37 -
Agus Suhartanto, - 132 -
Akha, - 110 -
Al Jazeera, - 121 -
aliceae, - 71 -
Alor, - 123 -
Alor Janggus, - 123 -
Amadpur, - 29 -
amruthae, - 40 -
ana, - 35 -
anaknesaeat, - 86 -
anamitica, - 96 -
ancestor worship, - 110 -, - 113 -, - 114 -
ancestral homes, - 52 -
ancestral shrine, - 114 -
Andaman, - 9 -, - 16 -, - 21 -, - 126 -
Andaman Sea, - 126 -
Andhra Pradesh, - 23 -
Angkor, - 77 -, - 83 -
Angkor Borei Archaeological Museum, - 78 -
Angkor Wat, - 79 -
animism, - 110 -, - 130 -
Antpur Radhagovindjiu, - 31 -
apicalis, - 60 -, - 73 -, - 96 -
Apis, - 18 -, - 101 -
Arab, - 126 -
Arabic influence, - 126 -
arakki, - 17 -
Archaeology, - 33 -
architecture, - 97 -
arcifera, - 15 -, - 36 -
Arunachal Pradesh, - 25 -, - 26 -
Ashish Kumar Jha, - 25 -
ashishi, - 25 -
Assam, - 9 -, - 23 -, - 26 -, - 34 -, - 35 -, - 38 -
At-chala, - 30 -
atripes, - 59 -
attic, - 123 -, - 156 -, - 157 -
Azad Jammu, - 12 -
Aztec culture, - 138 -

B

Bajaree Chuttong, - 76 -
Bamboo hives, - 53 -
Bandhavgarh National Park, - 28 -
Bangkok, - 52 -
Bangla, - 29 -, - 30 -
Bangladesh, - 9 -, - 22 -, - 23 -, - 28 -, - 29 -, - 32 -, - 33 -, - 35 -, - 134 -, - 173 -
Bardebhata, - 25 -
Bardhaman, - 32 -
Barrel Vaulted, - 153 -, - 154 -
Batanes Island, - 115 -
Bathudi, - 24 -
Battle of Hastings, - 38 -
Beescape, - 132 -
beeswax, *- 18 -*
Belawing, - 132 -
Bell Roof, - 153 -, - 154 -
bengalensis, - 15 -, - 35 -
Bengali, - 29 -, - 32 -, - 33 -
Bengaluru, - 16 -
Benguet, - 147 -
Betang, - 140 -
Betang House, - 140 -
Betotvear, - 85 -
Bhubaneswar, - 22 -
Bhutan, - 9 -, - 36 -, - 38 -, - 42 -
Bihar, - 23 -, - 28 -
Bima, - 130 -
binghami, - 60 -
Binh Thuan, - 92 -
Bishnupur, - 29 -, - 30 -, - 31 -, - 32 -
black netting, - 12 -
bonnet roof, - 129 -
Borneo, - 130 -, - 131 -, - 133 -
Borobudur, - 130 -
box, - 102 -
Box Gable Roof, - 153 -, - 154 -
box hives, - 52 -, - 53 -, - 126 -, - 132 -
Brahmaputra, - 35 -
brains and fluids, - 134 -
British colonialists, - 133 -
British Raj, - 40 -
Buddhism, - 52 -, - 55 -, - 130 -
Buddhist, - 37 -, - 38 -, - 53 -, - 55 -
Burmese, - 55 -
burrowing, - 135 -
Butterfly roof, - 11 -
Butterfly Roof, - 153 -, - 154 -

C

cacciae, - 15 -, - 28 -
calophyllae, - 15 -, - 20 -
Calophyllum inophyllum, - 20 -
Cambodia, - 52 -, - 53 -, - 77 -, - 78 -, - 79 -, - 81 -, - 83 -, - 101 -, - 109 -, - 118 -, - 120 -
cambodiensis, - 77 -, - 96 -
Cameron, - 17 -, - 35 -
camphor, - 134 -
Cangnan, - 113 -
canifrons, - 59 -
canopy, - 156 -
carpenteri, - 96 -
carrion bees, - 135 -
Ceylon, - 15 -, - 44 -
Cham, - 90 -, - 91 -
Champa, - 83 -
chandrai, - 15 -, - 21 -
Changlang, - 26 -
Char-chala, - 30 -
Charpoys, - 24 -
Chea Samath, - 77 -
Cherthneecha, - 17 -
Chhattisgarh, - 22 -, - 24 -, - 25 -
Chhotanagpur Plateau, - 22 -
Chiahseh, - 85 -
Chiang Mai, - 52 -
Chiang Rai, - 110 -
China, - 48 -, - 53 -, - 55 -, - 78 -, - 113 -, - 120 -, - 122 -, - 126 -
Chinese architecture, - 55 -
Chinese cemetery, - 135 -
chofah, - 79 -
Chofah, - 52 -
churas, - 31 -
Clerestory, - 153 -, - 155 -
Climate Change, - 121 -
coastal, - 127 -
coatepantli, - 138 -
Cochin, - 21 -
Cockerell, - 17 -, - 36 -, - 77 -
Codex Tovar, - 138 -
cogon, - 115 -
cogon grass, - 115 -
collina, - 59 -, - 77 -, - 96 -, - 172 -
Combination Roof, - 153 -, - 155 -
Comfort Indicators, - 71 -
Con Cuong, - 120 -
concrete anchors, - 125 -
Conical, - 155 -
Conical Roof, - 153 -, - 155 -
conical roofs, - 155 -
Contemporary House, - 70 -
crawling insects, - 125 -
Cross-Hipped Roof, - 153 -, - 155 -
Curved Roof, - 153 -, - 155 -

D

Dalan, - 31 -, - 32 -
Dambaul Chraen, - 85 -
dammar bee, - 101 -
Dara Sin, - 85 -

Dayak, - 131 -, - 132 -
Dayaknese, - 140 -
Dayaks, - 131 -
Dehong Dai, - 26 -, - 55 -
Dhangad, - 22 -
Dhangar, - 22 -
Dhemaji, - 35 -
Dhimar, - 28 -
Dibang Valley, - 26 -
Diknagar, - 29 -
Dipterocarp, - 17 -
Disco sphere, - 123 -
Do-chala, - 30 -
doipaensis, - 72 -
Dome Roof, - 153 -, - 156 -
Domed Vault Roof, - 153 -, - 156 -
Donyi-Polo, - 25 -
Dormer Roof, - 153 -, - 156 -
Dr. Le Duy Dai, - 90 -
Dr. Sweta Chakraborty, - 120 -
dragon, - 53 -
dragon ornament, - 113 -
Dravidian, - 22 -
Dropped Eaves, - 153 -, - 156 -
Dutch, - 128 -, - 129 -, - 156 -, - 157 -, - 158 -
Dutch Bonnet, - 129 -
Dutch Gable, - 156 -
Dutch Gable Roof, - 153 -
Dutch influence, - 130 -
DW News, - 120 -
Dzongu, - 36 -

E

Ecozone, - 36 -
eggs, - 97 -
ek-bangla, - 30 -
ek-ratna, - 31 -
England, - 38 -
Entomology, - 109 -
Europe, - 120 -

F

fascia, - 52 -, - 54 -
Fascia, - 55 -
Faseeh, - 15 -, - 20 -
fimbriata, - 72 -
finial, - 11 -, - 110 -
finials, - 52 -, - 53 -, - 54 -, - 79 -
Fishmongers' House, - 86 -
Flat Roof, - 153 -, - 157 -
flavibasis, - 72 -
Flavotetragonula, - 15 -, - 20 -
Flores, - 130 -
foyer-like, - 128 -
Freddie Lozada, - 124 -
frogs, - 132 -

Funan Era, - 77 -
furva, - 72 -
fusco-balteata, - 72 -
Fusion, - 126 -

G

Gable, - 52 -, - 128 -, - 153 -, - 157 -
gable and hip roofs, - 54 -, - 155 -, - 159 -
Gable Roof, - 153 -, - 156 -, - 157 -
Gable/rake carvings, - 52 -
Gadaba, - 23 -
gajashimha, - 79 -
Gambrel Roof, - 153 -, - 157 -
Gangtok, - 36 -
Garo, - 134 -
Garuda, - 52 -
Gawahon Eco Park, - 124 -
gazebos, - 156 -, - 157 -
Geniotrigona, - 59 -
Goaribari, - 133 -
Goddess Atite, - 78 -
Golaghat, - 26 -
Governor's Mansion, - 128 -
graveyard, - 135 -
Greek, - 27 -
green roof, - 159 -
gressitti, - 15 -, - 26 -, - 96 -
Grey Shutters, - 85 -
ground beams, - 125 -
Gulf of Thailand, - 126 -
Gunung Ledang, - 135 -
Gutob, - 23 -

H

Harimanok, - 53 -
harvesting, - 102 -
Head Hunter, - 131 -
headhunters' domicile, - 136 -
Heat Wave, - 119 -
heavy rains, - 159 -
hemileuca, - 73 -, - 96 -
heritage, - 132 -
Hexagonal, - 128 -, - 157 -
Hexagonal Roof, - 153 -, - 157 -
High House, - 86 -
hilts, - 132 -
Himalayan States, - 34 -
Hindu, - 49 -, - 78 -, - 93 -
Hinduism, - 130 -
Hip, - 128 -, - 157 -, - 158 -
Hip and Valley Roof, - 153 -, - 157 -
Hip Roof, - 153 -, - 158 -, - 159 -
hirashimai, - 72 -
Ho Chi Minh City, - 97 -
Ho Chih Min city, - 53 -
Hobby Horse, - 85 -

Hokkien, - 113 -
Hollander, - 123 -
Homotrigona, - 59 -, - 71 -, - 72 -, - 96 -
Honey, - 11 -, - 18 -, - 19 -, - 172 -, - 174 -
Honey production, - 19 -
Hooghly, - 31 -, - 35 -
horizontal cladding, - 85 -
Hoshangabad, - 28 -
Houaphanh, - 110 -
Huitzilopochtli, - 138 -
Hunli, - 26 -
hydroelectric dams, - 23 -

I

Iban tribe, - 131 -
Igorot, - 133 -, - 136 -
India, - 19 -, - 35 -, - 49 -, - 52 -, - 101 -, - 102 -, - 120 -, - 121 -, - 134 -, - 136 -
Indian Influence, - 126 -
indigenous, - 130 -
indigenous tribes, - 130 -
Indo-China, - 52 -, - 53 -
Indo-Malaya, - 11 -, - 131 -, - 135 -
Indonesian archipelago, - 130 -
Indra, - 139 -
insect tourism, - 52 -
Insect Tourism, - 126 -
iridipennis, - 15 -, - 44 -, - 59 -, - 102 -, - 172 -, - 174 -
Isaan, - 69 -
Isan, - 79 -
Islam, - 49 -, - 92 -, - 130 -
Islamabad, - 11 -
Islamic Architecture, - 30 -
Island Traditional Houses, - 123 -, - 128 -
Istana Jahar Palace, - 128 -
Istana Ulu, - 129 -
Isthmus of Kra, - 130 -
itama, - 139 -
Ivatans, - 115 -

J

James Brooke, - 131 -
Japan, - 122 -, - 142 -
Japanese, - 114 -
Japanese aesthetics, - 114 -
Java, - 52 -, - 53 -, - 130 -
Java Sea, - 126 -
Javanese, - 114 -
Javanese Joglo, - 114 -
Jayavarman VII, - 83 -
Jerkinhead, - 153 -, - 158 -
Jharkhand, - 22 -, - 23 -

Jinxiang, - 113 -
Jorhat, - 26 -

K

Kachin, - 26 -
Kachin Hills, - 26 -
Kadazan, - 134 -
Kadazan-Dusun, - 133 -, - 134 -
Kalimantan, - 131 -, - 132 -
Kamrup, - 35 -
Kandy, - 44 -
Karnataka, - 17 -
Kashmir, - 12 -
Kathodi, - 27 -
Katkari, - 27 -
Kdab Phtek, - 87 -
Kedah, - 123 -
kee suit, - 101 -
Kelantan, - 135 -
Kenyah Dayak, - 131 -
Kerala, - 17 -, - 19 -, - 102 -
Keshta Raya, - 29 -
Khalid Ali Khan, - 11 -
Khasi, - 35 -
    language, - 17 -
khasiana, - 35 -
Khmer, - 77 -, - 78 -, - 79 -, - 80 -, - 81 -, - 82 -, - 83 -, - 86 -, - 87 -, - 88 -, - 101 -, - 130 -
Khmer empire, - 79 -
Khmer influence, - 130 -
Khond, - 24 -
khpasa, - 86 -
Kimouch Seng, - 85 -
King Faisal Mosque, - 11 -
King Sovanakoad, - 78 -
King Suryavarman II, - 79 -
Kisan, - 22 -
Kolkata, - 35 -
Kompong Chèn Tbon, - 85 -
Kompong Thom, - 85 -
Konyak tribe, - 134 -
KT Chan, - 123 -
Kuala Kangsar, - 129 -
Kurukh, - 22 -
Kusong, - 36 -
Kutai Dayak, - 131 -
Kyrdemkulai, - 35 -, - 36 -
kyrdemkulaiensis, - 15 -, - 36 -

L

Labuan Municipality, - 134 -
laeviceps, - 15 -, - 59 -
Lakhimpur, - 35 -
Lâm Đồng, - 26 -
Langkawi, - 126 -, - 127 -
Langkawi Cultural Centre, - 126 -

Lanna, - 79 -
Lao, - 101 -
Laos, - 53 -, - 110 -, - 130 -
large spiders, - 132 -
Le coq, - 53 -
Lepidotrigona, - 15 -, - 17 -, - 36 -, - 59 -
Lepidotrigona arcifera, - 17 -
Lepidotrigona ventralis, - 97 -
limestone walls, - 115 -
Lingthem, - 36 -
Lingzya, - 36 -
Lisotrigona, - 15 -, - 21 -, - 27 -, - 28 -, - 72 -, - 96 -, - 97 -, - 172 -
Lisotrigona carpenteri, - 97 -
lizards, - 132 -
Lohit, - 26 -
Loloq', - 132 -
Lombok, - 129 -, - 130 -
London, - 121 -
Lophotrigona, - 59 -
Lumbung, - 129 -
Lushai, - 39 -, - 40 -
lutea, - 59 -, - 72 -

M

Madhya Pradesh, - 22 -, - 28 -
Madurese, - 139 -
Mae Hong, - 119 -
Maharashtra, - 9 -, - 16 -, - 22 -, - 27 -
Majapahit, - 52 -, - 130 -
Makassarese, - 130 -
Malaccan, - 129 -
malaipanae, - 73 -
Malancha Dakshina Kali, - 31 -
Malay, - 101 -
Malay architecture, - 130 -
Malay cemetery, - 135 -
Malayness, - 130 -
Malays, - 126 -
Malaysia, - 113 -, - 123 -, - 126 -, - 128 -, - 129 -
Malaysian Architecture, - 128 -
males, - 97 -
Mallaah, - 28 -
Maluku, - 130 -
Mamley, - 37 -
Manchar, - 12 -
Mandalay, - 119 -
Mandau, - 132 -
Mankalale, - 24 -
Mansard Roof, - 153 -, - 158 -
Mantam, - 36 -
Mao Heng, - 86 -
Maranao, - 53 -
Mardhari, - 28 -
maroam, - 101 -

marvel idea, - 124 -
medicinal, - 18 -
Medicinal properties, - 18 -
Meghalaya, - 34 -, - 35 -
melanoleuca, - 73 -
Melipona, - 35 -
Meliponines, - 97 -
MGVI, - 135 -
Michael Trané, - 77 -
millennials, - 123 -
Min language, - 113 -
minaret, - 11 -
Mindanao, - 130 -
Miri, - 25 -
Misri Jenu, - 17 -
Mizo, - 39 -, - 134 -
Mizoram, - 34 -, - 39 -, - 40 -
moats, - 125 -
mohandasi, - 15 -
Mohd Noor Isa, - 135 -
Momin Mosque, - 33 -
Monitor, - 158 -
monitor roof, - 158 -
Monitor Roof, - 153 -, - 158 -
Moorish style arches, - 126 -
Moors, - 128 -
Morigaon, - 35 -
Moro Malay influence, - 130 -
Mosques, - 76 -
M-Shaped Roof, - 153 -, - 158 -
MSNBC, - 120 -
Mư dâu, - 91 -
Mud house, - 47 -
Mughal, - 32 -, - 33 -
Munda, - 22 -, - 23 -
Murut, - 133 -, - 134 -
Museum, - 128 -, - 132 -
Museum of Ethnology, - 90 -, - 91 -
Muslim Beekeepers, - 11 -
Muslim Restaurant, - 76 -
Myanmar, - 53 -, - 56 -, - 119 -, - 130 -, - 134 -
Myinmu, - 119 -

N

Naga, - 52 -, - 134 -
Nagaland, - 9 -, - 34 -, - 36 -, - 37 -, - 38 -, - 39 -, - 134 -, - 136 -
Nagaon, - 35 -
Nagesia, - 22 -
Nagpur, - 25 -
Nasiruddin, - 126 -
necrophagous, - 135 -
nectar, - 19 -, - 97 -
Negros, - 124 -
Neil deGrasse Tyson, - 122 -

Nepal, -9-, -13-, -17-, -28-, -36-, -38-, -42-, -172-
ngaphamang, -17-
ngapkhyndew, -17-
ngapsiwor, -17-
Nghe An, -120-
Nicobar, -9-, -16-, -21-
Ninh Thuan, -90-, -91-, -92-
Nishad, -28-
Nomen nudum, -35-
None tribe, -135-
Nordic, -38-
NTB, -130-
NTT, -130-
nuptial flight, -102-
Nusantara, -131-

O

Odisha, -22-, -23-, -24-
ong, -101-
Orang Ulu, -131-, -136-
Oraon, -22-
Out of the box, -123-

P

pagdeni, -72-
pagdeniformis, -72-, -73-
pagodas, -37-, -52-, -53-
Pakistan, -1-, -9-, -11-, -12-, -28-, -120-, -121-
palaces, -53-
Palawan, -130-
Papua, -133-, -136-
Papua New Guinea, -133-
Pashighat, -26-
patahu, -140-
pediments, -11-
Penang Island, -135-
Penggilingan, -139-
Penggilingan Padi, -139-
peninsularis, -96-
Perak, -129-
perlucipinnae, -15-, -20-
Perspex, -12-
Peru, -37-
pet owl, -132-
phat moc, -92-
Philippines, -1-, -52-, -115-, -133-, -136-
Phimai, -83-
Phteah-Kataing, -77-
Phteah-Khmer, -77-
Phteah-Pit, -77-
Phteah-Ruang-Doeung, -77-
Phteah-Ruang-Dual, -77-
Phuket City, -128-
Pnrbrapheh, -85-

pollen, -19-
polygamous, -24-
Polynesian, -130-
Polynesian islands, -130-
porch, -128-, -155-
Portuguese, -51-, -128-, -129-, -130-
Portuguese influence, -128-, -130-
praeterita, -15-, -44-, -59-
Propolis, -18-, -19-
psychedelics, -123-
Pteas Khmer, -78-, -82-
Pteas Koeng, -78-, -82-
Pteas Kontaing, -78-, -81-
Pteas Pit, -78-, -80-
Pteas Rongdeung, -78-, -81-
Pteas Rongdorl, -78-, -80-
Pugh, -35-
Punjab, -11-, -12-, -13-
Pushpal, -25-
Pyramid Hip, -153-, -159-
pyramidal, -114-

Q

queens, -97-

R

Raghavesvara, -29-
Raghunatha, -32-
Raja Brooke, -131-
Raja Mangala, -52-
Rajasthan, -25-, -30-
rajithae, -40-
Rajput, -32-
Rajshahi, -23-
rake boards, -55-
Rangpur, -23-
Rarh, -33-
Rasmancha, -32-
Rasmussen, -26-, -35-, -44-, -77-, -173-
Rate of failure, -97-
Ratna, -31-
rats, -132-
resin, -17-
revanai, -15-, -27-
rhythm and flow, -114-
Ri-Bhoi, -34-, -35-
ridge-post framing, -86-
riots, -139-, -140-
rivers, -140-
Rojeet, -15-, -35-
roof ornaments, -52-, -53-, -54-
roof pitch, -54-
rotting meat, -135-
Roubik, -135-

Royal University of Fine Arts, -77-
ruficornis, -15-, -28-, -59-
Rungyung Chu, -36-
Russey, -87-

S

Sabah, -133-, -134-
Sagada, -147-
Sajan Jose, -15-, -21-, -27-
Sakagami, -15-, -26-, -174-
Sakyong-Pentong, -36-
Salem, -24-
Saltbox Roof, -153-, -159-
Samarinda, -132-
Sampit, -139-
Sandung, -140-
Sang Voeuy, -86-
Santal, -23-
Santhal, -23-
Sarawak, -48-, -128-, -131-, -133-, -136-
Sarimanok, -53-
Sawantawadi, -27-
Sawtooth Roof, -153-, -159-
scissors, -110-
Sema Naga, -37-
serpent, -53-, -138-
Seth Williams, -160-
Shades of Meliponiculture, -101-
Shanas, -15-, -20-
shandies, -24-
Shed Roof, -153-, -159-
Shishira, -25-
shishirae, -25-
shoebox cuboid, -52-
Shubham Rao, -25-
shubhami, -25-
Sibsagar, -35-
Siem Reap, -80-, -85-
Sikkim, -17-, -36-
sikkimensis, -37-
Sindh, -12-
Singapore, -113-, -114-, -128-
SINGAPORE, -59-
Singpho, -26-
Sinhala, -44-
Sinhala Janathāva, -44-
Sino Portuguese, -128-
sirindhornae, -73-
Siva, -29-
Sivasagar, -26-
Skillion Roof, -153-, -159-
small snakes, -132-
Smith, -15-, -17-, -28-, -44-, -59-, -60-, -77-, -172-, -174-
snail horns, -38-
Sora, -23-, -24-

South Asia, - 121 -
South China Sea, - 126 -
Spaniards, - 115 -, - 133 -
Spanish influence, - 130 -
spire, - 52 -
Sri Lanka, - 43 -
srikantanathi, - 15 -
Srü, - 86 -
St Bryce day massacre, - 38 -
stairway, - 127 -
stilts, - 127 -
stone houses, - 115 -
Straits of Malacca, - 126 -
strong gales, - 125 -
Stupas, - 38 -, - 52 -
sugarcane, - 114 -
Sultanate of Bima, - 130 -
Sultanate-era, - 33 -
Sulu, - 126 -, - 130 -
Sulu Sea, - 126 -
Suma, - 24 -
sumae, - 24 -
Sumbawa, - 130 -
Sümi, - 37 -
swallows, - 132 -
swiftlets, - 132 -

T

T. necrophaga, - 135 -
Ta Prom, - 85 -
Tadvi Bhils, - 27 -
Tagalog, - 53 -
Taiwan, - 26 -, - 115 -
Tamil Nadu, - 9 -, - 20 -, - 21 -
Taoist priests, - 114 -
Tausug, - 52 -
tectonic plates, - 12 -
Tenggarong, - 132 -
tenhaku - ki, - 38 -
Tenochtitlan, - 138 -
terminal passage, - 115 -
terminata, - 59 -
Tetragonilla, - 59 -, - 77 -, - 96 -
Tetragonula, - 15 -, - 17 -, - 20 -, - 24 -, - 25 -, - 26 -, - 28 -, - 35 -, - 36 -, - 44 -, - 59 -, - 72 -, - 73 -, - 96 -, - 97 -, - 174 -

Tetragonula bengalensis, - 17 -
Tetragonula iridipennis, - 17 -
Tetragonula laeviceps, - 97 -
Tetrigona, - 60 -, - 73 -, - 96 -
Texas, - 120 -, - 122 -
Thai Influence, - 126 -
Thai south, - 101 -
Thailand, - 48 -, - 49 -, - 52 -, - 69 -, - 70 -, - 71 -, - 76 -, - 83 -, - 112 -, - 119 -, - 130 -
Thang dơ, - 91 -
Thang tôn, - 91 -
Thenzawl, - 40 -
thenzawlensis, - 40 -
Theonym, - 27 -
Theravada, - 44 -, - 48 -
thermal comfort, - 71 -
thoracica, - 59 -
Thuan Thua, - 97 -
Tibetan styles, - 37 -
Timor, - 135 -
Timor Leste, - 135 -
Tingvong, - 36 -
Tlaloc, - 138 -
Toltec, - 138 -
totem, - 132 -, - 134 -, - 140 -
Tourist attraction, - 124 -
traditional dwellings, - 53 -
transparent wing, - 20 -
Travancore, - 20 -, - 21 -
travancorica, - 15 -, - 20 -
Trigona, - 101 -, - 102 -
Trigona crassipes, - 135 -
Trigona hypogea, - 135 -
Tripura, - 22 -, - 35 -
tzompantli, - 138 -

U

U.S. hegemony, - 133 -
Udaipur, - 25 -
UK, - 120 -, - 121 -
upsweep, - 53 -, - 113 -
upturned crescent, - 11 -
Uttar Pradesh, - 28 -
Uttaradit, - 119 -

V

Valley, - 157 -
valleys, - 157 -
vanes, - 123 -
Vat Phu, - 83 -
ventilation, - 70 -, - 78 -, - 119 -, - 157 -
ventralis, - 36 -, - 59 -, - 72 -, - 96 -, - 97 -
Victorias, - 124 -
Vietnam, - 53 -, - 83 -, - 90 -, - 91 -, - 97 -, - 112 -, - 113 -, - 120 -, - 130 -
VIETNAM, - 26 -
Vikram, - 24 -
vikrami, - 24 -
vines, - 140 -
Viraktamath, - 15 -, - 21 -, - 24 -, - 25 -, - 27 -, - 34 -, - 35 -, - 37 -, - 40 -, - 172 -
vulture bees, - 135 -

W

Wada, - 16 -
Walker, - 44 -
Wat Svay, - 85 -
wax, - 19 -
West Bengal, - 22 -, - 23 -, - 35 -, - 36 -
White Rajah, - 131 -
windmill, - 123 -
WION, - 122 -
withstand strong winds, - 115 -

X

Xiazetang, - 113 -

Y

Yeang, - 87 -

Z

zahv, - 110 -
Zhang Hui, - 114 -
Zhejiang, - 113 -
Zhou Daguan, - 78 -, - 79 -

Bibliography

Attasopa, K., et al.,. (2018). *A new species of Lepidotrigona (Hymenoptera: Apidae) from Thailand with the description of males of L. flavibasis and L. doipaensis and comments on asymmetrical genitalia in bees.* Retrieved from Zootaxa 4442 (1): 063–082: https://doi.org/10.11646/zootaxa.4442.1.3

Appanah(eds), S., & Turnbull(eds), J. M. (1998). A Review of Dipterocarps: Taxonomy, Ecology and Silviculture. *FRIM*.

Bista, S., & Shivkoti, G. (2001). Honey bee Flora at Kabre, Dolakha district. *Nepal Agricultural Research Journal 4-5*, 18-25.

Boonthai, & Sawatthum., A. (2014). Efficacy of stingless bee as Insect pollinator of rambutan var. Sri Tong. *MSc. Thesis. Rajamangala University of Technology.*

Cortopassi-Laurino, M., et al. (2006). Global meliponiculture: challenges and opportunities. *Apidologie 37*, 275–292.

Danaraddi, C. S., Viraktamath, S., & Bhat, K. B. (2009). Nesting habits and nest structure of stingless bee, Trigona iridipennis Smith at Dharwad, Karnataka. *Karnataka J. Agric. Sci., 22(2):*, (310-313).

D ash B.& Walia A. (2020). Role of multi-purpose cyclone shelters in India: Last mile or neighbourhood evacuation. ScienceDirect - Tropical Cyclone Research and Review Volume 9, Issue 4.

Das, P., et al. (2019). *Traditional bamboo houses of North-Eastern Region: A field study of Assam & Mizoram.* Retrieved from Key Engineering Materials Vol. 517 (2012) pp 197-202: doi:10.4028/www.scientific.net/KEM.517.197

Dejtisakdi, W., et al. (n.d.). Melliferous plants for the stingless bees (*Trigona collina* Smith) in the deciduous forest of Queen Sirikit Botanic Garden, Mae Rim, Chiang Mai Province. *Department of Botany, Faculty of Science, Kasetsart University*.

Eltz, T., et al. (2003). Nesting and Nest trees of stingless bees (Apidae; Meliponini) in lowland dipterocarp forest in Sabah, Malaysia, with implications for forest management. *Forest Ecology and Management 172*.

Engel, M. S. (2000). A Review of the Indo-Malayan Meliponine Genus Lisotrigona with Two New Species (Hymenoptera, Apidae). *American Museum Of Natural History.*

Iyengar, K. (2015). *Sustainable Architectural Design.* Routledge.

Jalil, A. (2014). *Beescape for Meliponines.* Singapore: Partridge Publishing.

Jalil, A., & Roubik(ed), D. (2016). Handbook of Meliponiculture. *Akademi Kelulut Malaysia.*

Jalil, A., & Roubik(ed), D. (2018). Handbook of Meliponiculture Vol 2. *Akademi Kelulut Malaysia.*

Karthick, K. S. et al. (2018). *Prospects and challenges in Meliponiculture in India.* Retrieved from International Journal of Research Studies in Zoology Volume 4, Issue 1, 2018, PP 29-38: http://dx.doi.org/10.20431/2454-941X.0401005

Lalhmunmawia & Das, S.K. (2019). *Social Structure of Mizo Village: a Participatory Rural Appraisal.* Retrieved from International Journal of Bio-resource and Stress Management 2019, 10(1):077-080: HTTPS://DOI.ORG/10.23910/IJBSM/2019.10.1.1899

Leonhardt, S. D., et al.. (2010). Stingless Bees Use Terpenes as Olfactory Cues to Find Resin Sources. *Chem. Senses 35*, 603–611.

Macías-Macías J. O., et al. (2011). Comparative temperature tolerance in stingless bee species from tropical highlands and lowlands of Mexico and implications for their conservation (Hymenoptera: Apidae: Meliponini). *Apidologie 42*, 679–689.

Nidup, T. (2021, April 26). *Report on the stingless bees of Bhutan (Hymenoptera: Apidae: Meliponini).* Retrieved from Journal of Threatened Taxa | Vol. 13 | No. 5 | Pages: 18344–18348: DOI: 10.11609/jott.4504.13.5.18344-18348

Ntawuzumunsi E., Kumaran S. & Sibomana L. . (2021). Self-Powered Smart Beehive Monitoring and Control System (SBMaCS) †. *Sensors, 21*, 3522.

Palittin D. & Hallatu T.G.R. (2019). Sar: Kanume tribal culture in environmental conservation to reduce global warming effects. *IOP Conf. Ser.: Earth Environ. Sci.* 235 012062.

R. Kajobe & C. M. Echazarreta. (2005). Temporal resource partitioning and climatological influences on colony flight and foraging of stingless bees (Apidae; Meliponini) in Ugandan tropical forests. *African Journal of Ecology*, Afr. J. Ecol., 43, 267–275.

Rahman1, A., Das, P. K., Rajkumari, P., Saikia, J., & Sharmah, D. (2015). Stingless Bees (Hymenoptera:Apidae: Meliponini): Diversity and Distribution in India. *International Journal of Science and Research (IJSR)*, Volume 4 Issue 1.

Rasmussen. (2008). Catalogue of the Indo-Malayan/Australasian stingless bees (Hymenoptera: Apidae: Meliponini).

Rasmussen, C. (2013). Stingless bees (Hymenoptera: Apidae: Meliponini) of the Indian subcontinent: Diversity, taxonomy and current status of knowledge. *Zootaxa 3647 (3): 401–428*.

Ratree S. M., Farah N., & Shadat M. S. (2020 DOI: 10.38027/N212020ICCAUA316262). Vernacular Architecture of South Asia: Exploring Passive Design Strategies of Traditional Houses in Warm Humid Climate of Bangladesh and Sri Lanka. *ICCAUA2020 Conference Proceedings.* Alanya, Turkey: AHEP University.

Rathor, V. S., Rasmussen, C. & Saini M. S. . (2013). New record of the stingless bee Tetragonula gressitti from India (Hymenoptera: Apidae: Meliponini). *Journal of Melittology No. 7, pp. , 1–5*.

Rattanawannee A., & Duangphakdee O. (2019). Southeast Asian Meliponiculture for Sustainable Livelihood. DOI: http://dx.doi.org/10.5772/intechopen.90344.

Reyes-González A., et al. . (2014). Diversity, local knowledge and use of stingless bees (Apidae: Meliponini) in the municipality of Nocupétaro, Michoacan, Mexico. *Journal of Ethnobiology and Ethnomedicine*, 10:47.

Roubik, D. (1995). Pollination of cultivated plants in the tropics. *FAO Agric. Serv. Bull. 118, Rome.*

Roubik, D. (2006). Stingless bee nesting biology. *Apidologie 37*.

Sakagami, S. F. (1978). *Tetragonula* Stingless Bees of Continental Asia and Sri Lanka (Hymenoptera, Apidae). *Jour. Fac. Sci. Hokkaido Univ. Ser. VZ, 2001. 21(2).*

Schwarz, H. F. (1939). The Indo Malayan Species of Trigona. *Bulletin of AMNH.*

Sommeijer, M. (1999). Beekeeping with stingless bees: a new type of hive. *Bee World 80(2):* 70-79.

Vasanthakumar, S., Srinivasan, M. R., & Thakur, R. K. (2017). Pollination behaviour of stingless bees Tetragonula iridipennis Smith on mango inflorescence in south India. *AKM- IMS3C.*

Vijayakumar, K. & Jayaraj R. (2013). Geometric morphometry analysis of three species of stingless bees in India. *International Journal for Life Sciences and Educational Research Vol.1(2), July*, 91 - 95.

Vijayakumar, K. (2014). Nest and colony characters of Trigona (Lepidotrigona) ventralis var. arcifera Cockerell from North East India. *Asian Journal of Conservation Biology, Vol. 3 No. 1 July*, 90–93.

Viraktamath, S. & Roy, J. (2022). *Description of five new species of Tetragonula (Hymenoptera: Apidae: Meliponini) from India.* Retrieved from Biologia (2022) 77:1769–1793: https://doi.org/10.1007/s11756-022-01040-8

Viraktamath, S. & Thangjam R. (2022). *Description of four new species of Lepidotrigona (Hymenoptera: Apidae: Meliponini) from northeast India.* Retrieved from Zootaxa 5175 (1): 001–030: https://doi.org/10.11646/zootaxa.5175.1.1

Vit, P., Roubik, D. W., & Pedro, S. M. (2013). Pot Honey - A Legacy of Stingless Bees. *Springer*, DO - 10.1007/978-1-4614-4960-7.

Vit, P.; Pedro, S.R.M.; Roubik, D. (2018). Pot-Pollen in Stingless Bee Melittology. *Springer International Publishing* AG. DO - 10.1007/978-3-319-61839-5.

Wimalaweera A.P.S. (2001). *Sri Lankan Vernacular Architecture As an Appropriate Response to The Basic Forces in Built Environment.* Sri Lanka: Faculty of Architecture, University of Moratuwa.

www.ingramcontent.com/pod-product-compliance
Lightning Source LLC
Chambersburg PA
CBHW061551010526
44117CB00022B/2986